DEBATE OF LIGHT AND DARK

- A 100 Year Bet with NASA

Dr. Yuxiang Wu

The most wonderful road is the one inside the Berkeley South Gate behind the author. It guides us to break directly through the blue sky.

Debate of Light and Dark

- A 100 Year Bet with NASA

Dr. Yuxiang Wu

Two-W Object
2016

Copyright Two-W Object
All rights reserved. No part of this publication can be reproduced, distributed or transmitted in any form of by any means, including photocopying, recording, digital scanning, or other electronic or mechanical methods, without the prior written permission of the publisher, except in the case of brief quotations embodied in critical reviews and certain other noncommercial uses permitted by copyright law. For permission requests, please address the publisher.

Debate of Light and Dark - A 100 Year Bet with NASA
Dr. Yuxiang Wu
Published 2017
Printed in the United States of America
ISBN-13: 978-1-68419-737-8
BISAC: Science / Cosmology -- Cosmology, cosmic model, Einstein, NASA, Big Bang, Olbers' paradox, dark matter, redshift, cosmic microwave background, whole sky map

For information address:

Yuxiang Wu, Two-W Object
yuxiangwu@outlook.com

Debate of Light and Dark - A 100 Year Bet with NASA
This book is published together with following versions in United States and Mainland China:
English book: <Debate of Light and Dark - A 100 Year Bet with NASA >
    ISBN: 978-1-68419-737-8, March, 2017
Chinese book: <光暗之争 - 与美国宇航局 (NASA) 的百年期约>
    ISBN of China: 978-1684197385, September 2016
    ISBN of USA: March, 2017, March, 2017
Due to different publication time, when there is content conflict in these versions, the contents of the English version shall prevail.

Dedicate this book to:

My wife Li Wang and my sons Eugene Wu and Johnny Wu, for their excellent work on the research, writing, thesis completing and translation for this book. My parents, brothers and sisters. My forever teachers, schoolmates, friends. Yichun, my root and home town. Berkeley, my second home town.

Special thanks to:

Thanks to the editors who selected the "Distance Mystery" (距离的奥妙) section of my book *"Who Should Talk about Cosmos (谁有权谈论宇宙)"* into the China national university 7.15 Collectibles language textbook. Without the encouragement of these editors, whom I have not met, I would not have had enough confidence and courage to finish this book.

Thanks to the editors of Matter Regularity magazine from the Beijing Relativity Association. Your long-term, unremitting efforts have made available and maintain a safe place for valuable research not accepted by the mainstream.

Thanks to the help from professional editor Jefferson of www.firstediting.com for the English version of this book.

TABLE OF CONTENTS

Debate of Light and Dark...i
Table of Contents..vi
Preface..1
Forward..9
Part i A 4th grade math quiz that has been fooling human a century..............17
    Chapter 1 SPECIAL RELATIVITY QUESTIONED..................................22
        1. The Contradictory results obtained while applying the Lorentz transformation to the relativity calculation for multiple reference systems.. 28
        2. "Moving meter stick is shorter" disaster...35
        3. "Moving clocks are slower" paradox..38
        4. The Error in Einstein's Paper Results From the Fact that the Mathematical Calculation is Correct but the Physical.............................54
        5. Why is the Relativity Principle Defined in Einstein's Paper Unreasonable? 72
        6. In full accordance with Einstein's method of proof, using the relativity principle defined in his paper to prove that the concept of relativity is not correct... 75
        7. Einstein's paper should not have extend the use of the local time as the whole systems time..76
        8. In the simple mathematics, there is profound truth- The philosophy in the mathematical limit theory says "no" to the principle of relativity................ 78
    Chapter 2 General Relativity Theory Questioned..87
        The proposition of "Gravitational field deflects light" proposition makes itself to be criticized.. 94
        Finding all factors which are bending the light that is going through near the Sun..98
        Design the experiment to look for all factors that affect light bending.........100
    Chapter 3 Discussion about Time and Space Concepts............................. 105
        Space is a physical objective reality... 105
        Time is not an entity, but a concept defined by human on the Earth, a ruler 106
    CHAPTER 4 The Expression and Application of Four-dimensional Space-time Diagram... 117
        Four-dimensional space-time diagram expression of the world line of the event object... 121

      Application of the Four-dimensional Space-time Object Event World Line Diagram 126

Part II The Big Bang Theory Criticism 135

CHAPTER 5 Tangle between Bright and Dark 137

An Overview of OLbers' Paradox 142

The Big Bang model as premise of Olbers' paradox solution 143

Status quo of Olbers' Paradox 148

Flaws of Olbers' Shell Layer model – New Solution by Distance 150

General Algorithm to Calculate night "dark" 159

Estimate of the Total Light Intensity from Extragalactic Stellar Irradiation on the Earth 162

CHAPTER 6 True Face of Hidden Celestial Objects and Dark Matter 169

Solution one of seeking: exploring the possibility of human migration out of the Earth 169

Solution two to find: if aliens exist, where are their bases? 176

Hidden Celestial Object Overview 184

Several fundamental factors affecting the ability to see celestial objects 186

Mathematical abstract definition of Limited Observable Radius 190

Definitions of Hidden Celestial Object (HCO) 191

The Hidden Celestial Object concept is more persuasive than the concept of dark matter 200

Chapter 7 Frequency rhythm of tint and ripple 207

From sound wave to light wave-- from the Doppler effect to Hubble's law. 209

If there is a difference between sound and color, why do we need a reinterpretation of celestial redshift? 210

The speed of light, the speed of celestial light, and the speed of celestial image 215

Telescope response time 221

The role of different regional Zones of batches of photons 225

Time Hysteresis Redshift and Displacement Redshift 234

CHAPTER 8 Dancing on a Sesame Seed 239

Questions about the Whole sky map of the cosmic microwave background 240

The light waves arbitrary passing through a point P in the space 247

A discussion of the isotropy of the microwave background with invisible wave source 249

The Information Promotion of NASA Contrary to Scientific Reasoning 254

  Seeing may be not real - universe dot matrix timed image - Further understanding of celestial images received.. 257
  Seeing may be not real - Is it a real planet or only a no source illusory light? 260
 CHAPTER 9 Criticism on the Big Bang Theory.. 265
  Introducing to the Big Bang.. 265
  The main theoretical basis of the Big Bang... 268
  NASA's negative effects.. 268

Part III Calls for the Creation.. 285
 Chapter 10 Foster the Creation... 287
  The purpose of education in modern society...................................... 287
  Modern education stifles the creation... 289
  The cost of creation... 292
  Inspiration obtained from the invention of Boolean algebra............. 295
  Discussion about genius — sorrow of modernized education........... 302
 Chapter 11 Mathematics is old.. 309
  Climbing trees and planting trees — Is Mathematics old?................ 309
  "End of Science" and the rising opportunities of science and technology... 314
  Solemn classical science and frivolous modern science.................... 315

Appendix: ON THE ELECTRODYNAMICS OF MOVING BODIES...... 319
Some Simple Prior Knowledge.. 326
Author Introduction.. 331
Ending Words.. 332

Reading Suggestion

  I had to write this book of scientific content in a way that the general public can understand. Though I tried my hardest to keep it interesting, there are some parts of the book where I am not very successful at this, lulling the reader into a drowse. Writing is imperfect, and there are always paragraphs or even entire sections I would like to rewrite to make more engaging. When you encounter such a section, please read ten lines at a glance, or directly skip the section, and continue on to the more interesting portions of the book.

PREFACE

The Discovery of Gravitational Waves Have in Mind
-Thoughts after Reading Yuxiang WU's "Debate of Light and Dark"

Dr. Qing Su (苏青[1])

In 1915, Einstein put forward his general theory of relativity, and the following year in February, in communication with the German physicist Karl Schwarzschild, he predicted the existence of gravitational waves. On February 11, 2016, David Reitze, head of the US LIGO (Laser Interferometer Gravitational-Wave Observatory, abbreviated as LIGO), announced that, with the two detectors located in Hanford, Washington, and Livingston, Louisiana, humans had directly detected gravitational waves for the first time. The relevant papers were published that day in the online edition of the American journal Physical Review Letters and the British journal Nature. Domestic and foreign media and the scientific community generally believed that the discovery of gravitational waves not only verified Einstein's general relativity predictions from 100 years ago, but also provided experimental evidence for the Big Bang theory and marked a milestone in physics and astronomy.

---

[1] Dr.Su, Qing, a researcher, Leader of Science Publishing house and China Science. Former "Degrees and Graduate Education" Editorial Board, deputy director of the president's office Beijing Institute of Technology, Director, Publishing House, China Association for Science and Society Service Center deputy director, etc. Leading talent of the Press and Publication Administration of Press and Publication industry.
http://baike.baidu.com/link?url=bWG5RbyQ_q3t6u5zUG3q2bGjkcNHOToKr_oVVCJQogvboB UceACCTgf09ltS4FT8ttrh1NnmBKhrNyaaXzLfOTjD1Wn9GOpXW9AfFsNbFh3

From now on, humans would have a new understanding of the universe.

At this moment, I am studying the academic science book Debate of Light and Dark written by Dr. Yuxiang Wu, a Chinese-American scientist, and I cannot help but think of two long-standing problems that have troubled me. First, can we doubt the major research achievements in the field of science? Second, who is eligible to question the major scientific achievements?

The first question address the most important nature of scientific research, but also provides an important prerequisite to carry out for scientific research and an important manifestation of scientific spirit. Therefore, the first question is not a problem and should not be a problem. However, when the owner of the results of scientific research is as famous as Einstein, when the results of scientific research institutions like LIGO of the United States have that authority, when those significant research achievements have been especially accepted by the scientific community, it may become a problem; in such cases, people tend to lose the courage to question, dispel the idea to challenge, and even accept without question the blind worship of these famous scientists and their serious scientific achievements.

Take this discovery of gravitational waves as an example of a major scientific achievement. The outcome was announced, the scientific community and, especially, the media were elated, aboil, and no one questioned the results. This reminds me of another scientific event that happened two years ago when the "primordial gravitational wave" was discovered. That was in March 2014, when a team of scientists from the Harvard-Smithsonian Center for Astrophysics announced the discovery of B-mode polarization signals in the cosmic microwave background radiation, which were likely to be imprints left by primordial gravitational waves. For a time, the media and the scientific community voiced one praise after another, calling the discovery of "original gravitational waves" a "significant achievement of Nobel Prize level." Unexpectedly, less than a year later, the team announced again that the discovery of the "primordial gravitational wave" was a scientific error – the observed signal from the interference of dust in the Milky Way rather than the original gravitational waves. Thus, not all of the major scientific discoveries or significant findings certainly are correct. We should encourage scientists to boldly challenge all claims, and media coverage should carefully hold sensible restraint.

I am very happy that Debate of Light and Dark is such a brave academic book to challenge major scientific research propositions such as the famous Olbers' Paradox, the idea that "light is deflected by gravitational fields," and the Big Bang theory. The book is divided into three parts: Big Bang theory criticism, Einstein's relativity criticism, and calls for the creation of a cosmological theory that specifically includes: no assumptions to solve Olbers' Paradox; defines the concept of hidden celestial objects based on the relative (absolute) limited observable radius; derives the mainly true cause of the celestial redshift by defining celestial image propagation velocity; examines NASA's cosmic microwave background whole sky map from a scientific perspective, comments on the use of NASA's COBE and other measurements, and shows how their model is unreasonable and unscientific; points out special relativity itself contains the "moving sticker shorter" disaster and the "moving clock slower" paradox; designs corresponding experiments for the verification of the disaster and paradox, and so on. For those experts and scholars in the relevant areas, the book might make them less complacent when there are other major scientific discoveries, or soberer and less arrogant when the birth of important research results from its ideas.

Mr. Yuxiang Wu is one of the first university graduates after the reopening of the College Entrance Examination. He is a professional graduate in underground coal mining from the Shandong Mining Institute, and has a master's degree in mine optimization design from the Beijing Graduate School of China University. After graduate school, he taught for a few years and then traveled to the United States to the University of California at Berkeley to pursue a doctorate in operations research in mining engineering. Dr. Wu has shown profound erudition and outstanding talent in a wide range of interests, but in his spare time, he is obsessed with cosmology. In his view, the universe is so wonderful and "the human understanding of the universe is a slow, sustained, and constant re-understanding of the process." Although Mr. Hawking once asserted: "we may now near the end of the search for the ultimate laws of nature," Mr. Wu believes that "not only do humans know little about the universe, but the knowledge already obtained is filled with questionable ideas. "Given the current level of technology, humans could spend even ten thousand years and still not cross the moat of one light year's distance," so Yuxiang Wu believes that for exploration of the deep universe, human beings can only passively accept the outside "light" (or "electromagnetic waves") to carry out cosmological research, and thus the present study of cosmology only consists of "passive waiting, daring speculation," four simple words that summarize the situation well. So, how to change the status quo? Dr. Wu says: "The first is to return to the basic spirit of science: to base research on facts; to have critical, groundbreaking ideas; and to dare to question authority predicated upon the basic scientific principles and introduce multi-party confirmation of new ideas. Remove the false and retain the true to smoothly study of the universe."

Despite his fame in the field of software development, Yuxiang Wu is, after all, not an astrophysicist. So when he questions such famous ideas as Olbers' Paradox, the "gravitational field deflecting the light," and major scientific research propositions like the Big Bang theory, people will inevitably doubt whether he has the qualifications and the ability to question such things. This is why I have been holding off writing the preface for Debate of Light and Dark, the most important reason for cautious attitude. Honestly, when I got the manuscript and saw the title page of tips, aphorisms, and the questioning of a number of important research results, my first reaction was, "This is a civil scientist's work." Like how, after the recently released discovery of gravitational waves, many in the media have dug up tape of an entertainment program from five years ago on Tianjin TV, in which a self-proclaimed "Brother-Nobel," laid-off worker Guoying Guo, mentioned the "gravitational wave" concept. Many have said that the guests on the show, including Zhouzi Fang, put collective "pressure" on Guo so that China lost a Nobel laureate scientist, and that the presenters and guests now need to apologize to Guo; but I know, despite the host's and guests' ridicule and sarcasm may have lacked respect for Guo, that Yingsen Guo is undoubtedly a typical "folk scientist." Guo, with only a junior-high-school education, is not the first person to propose "gravitational waves," and his "gravitational waves" concept was put forward only to explain that he had "discovered" a way to let cars run without wheels, and to put forth his "theory" that people can live forever. His idea does not have any relationship with the "gravitational waves" of physics and astronomy.

I have long served as head of academic journals and of a scientific publishing house, and I spend a lot of time each year, with much patience and hospitality, with contributors like Yingsen Guo, "folk scientists" who deny the theory of relativity, or have a proof for the Goldbach conjecture, or who've invented a perpetual motion machine, made some other major scientific breakthrough. The common feature of these people is that their education is generally low, they have not had rigorous scientific training, and they have paranoid personalities. During discussing problems, only he gushes narrative's sake, you don't have right to raise questions; usually he wants you to give judgment on his so-called "significant results" right away. He never wants to give the "significant results" document to you and let you send peer review. The reason is very simple, such a major scientific achievement, if reviewers would made the interception, and his devoting so "major achievement" Information obtained hardships, would have wasted for nothing?

Fortunately, after carefully reading the Debate of Light and Dark manuscript, I denied myself the same speculation on Dr. Yuxiang Wu. Wu not only received rigorous scientific training from a domestic first-class university and published astronomy and physics research papers in academic journals, but he was familiar with and tried to follow the basic paradigm of the scientific community. Debate of Light and Dark also shares the good-faith attitudes and expectations of the scientific community of exchange, discuss, learn, and contend. Although I am not an expert in astronomy or physics, I have received formal science and engineering training, study, and research from the university to graduate school, and coupled with Dr. Yuxiang Wu's superb control at the text level, extraordinary imagination, and genius of scientific communication, I can understand the general content of Debate of Light and Dark and accept the book's reasoning, argumentation, scientific experiments, and other methods, including even some studies in conclusion – despite the profound theoretical discussion of the major scientific issues. So, I think, Debate of Light and Dark has publishing value, and its content is also worth the effort for experts and scholars in related fields to discuss and then question.

In the 20th century, more and more scientific studies show the trend of cross-disciplinary integration, and astronomical research experts in the field do not have exclusive patents. In astronomy, more researchers are using mathematical tools to do things like calculate the trajectories of celestial bodies, to describe the marvelous sky above our heads, and to develop practical ways to meet the needs of people to determine things from observed phenomena, to obtain the correct time, to make calendars, and for other aspects of daily life. Ever since Galileo invented the telescope, human eyes have greatly extended while observing the sky; the use of radio telescopes, the Hubble Space Telescope, and other modern means of observation allow human beings to explore the depths of the universe. It is multidisciplinary scientists who continue to make celestial mechanics, astronomy, astrophysics, cosmology, astronomy, and other branches contribute to the rapid development of human understanding of the universe and the various types of astronomical objects and phenomena, helping that understanding to reach unprecedented depth and breadth. In this sense, although interdisciplinary, Dr. Yuxiang Wu is equally qualified and has the right to question such major scientific propositions as the famous Olbers' Paradox, "light deflecting by gravitational field," the Big Bang theory, etc.

Examples of interdisciplinary scientists making significant achievements in scientific research abound. The famous 19th century German chemist Friedrich A.

Kekule was a student of architecture early in his career. After changing to specializing in chemistry, he mainly engaged in the theoretical study of the structure of organic compounds, presented the first benzene ring structure theory, and greatly contributed to the progress and development of organic chemistry and the aromatic chemical industry. He also thought about constructing a three-dimensional atomic arrangement, the first time the concept of valence from the plane into three-dimensional space was universally recognized, and he has been a real authority in the organic chemistry community since the 19th century. This example may be a few years ago; we might need a more recent example! The 2003 Nobel Prize in Physiology or Medicine was awarded to American Paul Lauterbur and Peter Mansfield of UK in recognition of their groundbreaking achievements in the field of magnetic resonance imaging technology. Paul Lauterbur, a chemist, and Peter Mansfield, a physicist, both received the highest scientific honor in the medical field and the highest academic award – the Nobel Prize – side by side. This tells us that interdisciplinary research has not only not become an obstacle to scientific breakthroughs, but has provided the advantage of multi-disciplinary integration of innovation.

In fact, even fellow scientists in search of scientific truth are fallible and come a cropper. In 2006, the internationally renowned mathematician and academician Shing-Tung Yau declared that Zhongshan University Professor Xiping Zhu and American mathematician Huaidong Cao had thoroughly proved one of the most difficult mathematical problem of the Century – the Poincaré Conjecture. The truth is, as early as 2003, the Poincaré conjecture had been proved by Russian mathematician Grigori Perelman. Perelman thus also won the top prize of the annual international mathematical community – the Fields Medal; Finally, Zhu and Cao themselves even admitted that they did not make any new contributions to derive the Poincaré Conjecture. This is another way of saying that when scientists in different fields of carry out scientific research on a major question, even if an error has occurred, the scientific community should be more inclusive and tolerant.

However, it does not mean that I had no concerns about writing a preface for Debate of Light and Dark. In my opinion, Debate of Light and Dark is not in the strict sense an academic book. First of all, Mr. Yuxiang Wu's literary essays are a way to explore serious major scientific issues. According to his own words, the book is meant to follow "From philosophy to mathematical papers and then back to a literary description in the context of clear thinking." Therefore, although the writing of the book is beautiful and easy to understand, some of the reasoning and argumentation is inevitably filled with literary imagination and personal, emotional components, and it is difficult to ensure that it is rigorous and careful everywhere. Second, any scientific progress is based on previous studies. Debate of Light and Dark is based on the author's own published six papers as a reference, cited in the References but it does not do enough to cite previous studies, which makes the manuscript scientifically and academically play at a discount. Furthermore, I do not hold a positive attitude on all the explorations and final conclusions of the book. For example, in the argument on "gravitational field deflecting light" is a debatable proposition: the author points out that "the full light of the Sun can reach everywhere in space, so light is deflected to where?" And it takes this as an important basis for questioning the "gravitational field deflecting light" proposition. In fact, I think this sentence itself is debatable. Wind fills the entire space its movement can reach, but that does not mean that wind cannot be deflected. Because there are such defects and shortcomings, I temporarily name "Debate of Light and Dark" as popular scientific writing that places more emphasis on the dissemination of scientific knowledge, explore scientific questions, contend with the academic point of views, and challenge the active academic atmosphere, etc., to show the difference from a normal academic work.

This leads to another problem: are popular science academic works similar to Debate of Light and Dark worth publishing? I think, for natural science books, the publishing academic point of view does not mean that the book is correct, and it does not mean that the recommenders, the reviewers, those who write preface, or the readers, fully agree with the author's opinion; the published aim is to thus attract more attention of researchers and thinking about the major scientific issues the authors explore jointly, to participate in discussions, exchanges, so as to promote academic contention and scientific progress. After all, in today's China, we really lack the scientific questioning spirit, and there is a lack of encouragement and support for scientific questioning, academic contention, different points of view, and debate environments, also, we lack science warrior scholars like Dr. Yuxiang Wu, who have the courage to challenge the authority of science.

I am not only holding deep expectations on the publication of Dr. Yuxiang Wu's <Debate of Light and Dark>, but also hold profound respect for Dr. Yuxiang Wu.

Is that preface.

March 15, 2016, early morning in Willow Apartment, Haidian District

FORWARD

> My life is finite, but knowing is infinite. To run after the infinite with the finite is truly wrong!
> - Zhuangzi · The Inner Chapters · Health Master Chapter III

I often believe that sharp thinking
Able to pierce through all that is fuzzy and false
That sees through the deepest darkness of the universe
Must be the bright shining of our wisdom

But the distance is too far
The path to the heavens is infinitely long
Neither can we travel at the speed of light
Nor can we live long like the God

The view of the telescope will be blocked by the noise
Our exploring eyesight finally ends in a faraway emptiness
The universe leaves us forever searching through his impossible confusion
With his veil of unreachable distance

So we hold our hearts that eagle to explore the deepest parts of the universe
But stand solid and make our real dream true on Earth with limited budget

It took me 3 years to get my PhD from U. C. Berkeley. Now I am contributing the essence of my 10 years of thinking to you. I hope you will like it.

As long as you have at least a middle school education, and the patience, you can understand 99% of this book.

In addition, in the process of writing I deliberately incorporated the contents of the book and emphasizes how the thinking evolved from Elementary

Mathematics to Higher Mathematics, how to develop our own ability keen to discover and to solve problems, how to turn the practical examples into scientific research subject, how to develop our own creativity, how to use a simple elementary mathematics challenge the experts who use complex mathematical dreamlike models ... and such of the content that after you mastered then will be useful to you in your lifetime.

Let us work together to re-understand those important issues puzzling the contemporary physical world, such as the contradictions inherent in Einstein's original relativity theory and the Big Bang theory (including the so called most important task of physics of finding dark matter in the next few decades), and change the stagnation of scientific theory, innovation, and creation.

Together, we'll come to know that human understanding of the universe is a slow, sustained, and constantly renewing process. Let's think together how we can activate the theoretic creativity of human.

A very interesting example from recent events can give us some insight.

A friend sent me 100 ancient poems about spring, including "Birds Sing in Stream" written by Wang Wei: "Idle man drops osmanthus flower, on a silent night and empty spring hill. Moon rise shocks the hill birds, who sometimes sing in the spring stream."

I asked my friend: "The osmanthus flower is fragrant in autumn. It blooms in autumn and not the spring, right?"

My friend answered: "Spring Hill is the name of the hill, so the Osmanthus flower matches the poem."

I asked again: "Then what about the last sentence?"

My friend kept his silence.

This discussion, perhaps, can come to an end.

But still, I did not feel easy. The famous poem has a thousand years of tradition, and it is impossible that nobody had found this problem before.

So I searched in the internet.

First, I saw a paper explaining the poem. In the paper, the problem was noticed, and it was explained that "osmanthus" means "four season osmanthus." Oh, Wang Wei was correct.

This should have been the end of the story. But the next result of the internet search confused me again. In the encyclopedia entry on "four season osmanthus," it was explained that its flower blooms in the autumn.

The encyclopedia said the four season osmanthus blooms in the autumn, so

Wang Wei was wrong again.

What is the real and true story?

So I continued my search and reading and finally discovered that four season osmanthus is a variety of the osmanthus that is different from other osmanthus in that it blooms in every season of the year.

Now Wang Wei was correct again.

I went back and forth on this a few more times before finally putting together a few thoughts:

- My friend is a senior professor at a famous university. But while reading the famous poem, he never thought that there might be something wrong with it, even though it seems fairly obvious that the poem confuses spring and autumn. This is the common attitude taken while reading the works of famous authors. It can be called inertial thinking.

- It is possible that there are mistakes inside classic scientific books like an encyclopedia.

- We need to think carefully and to base our reasoning on common science and knowledge. The more surprising and unusual a thing, the more we need to use basic thinking in our reasoning about it.

It is the same with respect to our awareness of the problems of the universe.

For example, Mr. Stephen Hawking, who is an authority in astronomy, has asserted: "We may now be near the end of the search for the ultimate laws of nature." So ordinary people will follow him and believe that humans really have complete insight into the universe and know everything there is to know about heavenly things. But, in fact, this is a totally wrong impression. Humans know very little about the universe. Even existing knowledge is filled with questions and problems.

Ancient people observed the universe through their eyes only. If there is no Sun or moon for comparison, the Milky Way's starry expanse constitutes a rigid, two-dimensional plan. Although modern people can make various observations of certain parts of the universe using telescope-based instruments and can see with a deeper penetration, the way cosmologists explore the universe, in essence remains, the same as ordinary people.

The way mankind explores universal mysteries can be summed up in four words:

"Passively waiting, daring speculation."

These are very sad characters.

"Passively waiting" means that mankind can't fly across a distance of even one light year without using 10 thousand years! So, mankind is forced to stay on Earth and to study two kinds of objects coming from outer space. The first consist of small amounts of meteorite-like objects dropping on the Earth. The other is starlight, which consists of all kinds of light. According to their different frequencies, they can be called microwaves, electromagnetic waves, ultraviolet light, infrared light...

Meteorite-like objects can bring us some information, but their quantity is too little, they mostly represent a too small distance, and we often cannot determine their source. They cannot depict the face of the universe.

Light is the primary target for humans to study the universe.

It is impossible for humans to actively emit radar to explore the universe.

Humans can only wait passively for the arrival of light or electromagnetic waves from outer space. It is not that people do not want to take the initiative, it's that no initiative is possible. People don't have the slightest ability to proactively detect distant luminous objects. Coupled with slow speed of human travel, the distance that human beings need to cross to reach other bodies becomes unthinkably huge. Even the places cosmologists have called "the backyard garden of Earth," located just a few light years away, we are not able to reach. Humans can only hover in place in the vast space of the elusive star road.

Regardless of whatever news or reports name those space probes, satellite detectors, microwave detectors, and so on, we need to be clear: their "probing" or "detecting" merely refer to their "receiving" light for analysis. For example, the COBE satellite that detected the cosmic microwave background is actually "receiving" microwaves from space near the Earth.

Of course, astronomy experts can take the initiative to do something, such as make more advanced telescopes, align and focus their telescopes on certain luminous objects so that they can receive most of the lights emitting from it, or, in order to reduce noise, send the telescope outside of the atmosphere, etc. Doing these things can lower human-controllable noise a little bit, which we'll discuss later. For those uncontrollable factors, including natural noise, there is currently no way to improve observation ability by reducing them.

After receiving the information of the celestial light, cosmologists must use a variety of instruments to process the information, using various methods. Based on this analysis, cosmological models are made; the face of the universe is depicted. The impunity with which contemporary cosmologists often perform this step, breaking classical tradition and using daring speculation and imagination, really is amazing, using "daring speculation" to describe is not too much!

Due to the huge differences in control over resources, and the huge gap between the professional and the amateur, the vast majority of humanity does not have first-hand information about the universe. The stars in the metropolitan night sky are often disguised by city lights, and in the country, the dull sky cannot stimulate the villagers' idea that they are only seeing the same picture every night. Narrow specialization makes the public only listen to astronomy experts. People are convinced by the cosmological model that the experts depict, and give up any urge to explore the universe.

However, suppose that the cosmological masters, who have full right to speak, made mistakes that deviated their understanding from the right track? What would that mean?

For example, one topic discussed all the time is whether or not humans will eventually be able to move to planets outside the solar system.

According to the "bold speculation" of cosmological experts, and from their theoretical results, we do not see the slightest possibility of success. In other words, based on the cognitive approach of the experts, the issue of immigration out of the solar system is basically impossible.

Then, can we, based on the observational data of the experts, use new theories and ideas from a different perspective so as to inject some positive energy into this seemingly unsolvable problem? Can we also apply that energy to other puzzling cosmic problems to provide some new ideas for solving them?

Cosmological science believes in "WYSIWYG", what you see is what you get. The person whose telescope can see farther and more clearly, who grasps the voice, is the person whose data is authoritative data. But this is not correct!

In this book we will prove: the distant cosmic images we see are not the real face of the universe; we are often not seeing what's real! The image of the distant universe is depicted in units of years apart, a completely different lattice with ten thousand years as a time unit. It is depicting elusive, illusory beauty of the scene. People never glimpse through the cosmic suspended veil!

Therefore, the telescope, whether good or bad, is not sufficient to grasp the understanding of the universe, but it still became the basis for the right to speak the truth! But often, awareness of the distant universe interferes with and becomes the resistance to know the true face of the universe.

Unfortunately, many in the cosmology fields seem to have moved away from the right track, with many of them doing nothing but portraying the face of the universe as blurred illusion. Their studies tend to focus on one aspect while sometimes forgetting another. When these various aspects have finally been put together, the result has often been conflict. One typical example is the application of the Cosmological Principle. When applied to the measurement results of the microwave background, and when applied to the search for dark matter, the results are freakishly contradictory.

To change such situations in cosmological study, investigators should first return to the basic spirit of science. They need to have critical, groundbreaking ideas, to dare to question the judgments of authority based on basic scientific principles, and to have multi-confirmation before launching new ideas. In such a way, they can get rid of the false, keep the truth, and proceed with a broad, smooth research process for studying the universe.

Cosmological view dominates our view of the world. We need a correct worldview.

This is the mission that drives me to sit in front of the computer. Whether it is a sunny weekend or a lantern hanging festival, I am brooding, searching, hammering. Within this book is the landscape of my journey of reflection, the sudden realization while waking up from a midnight dream...

To support the arguments of the book, I have designed four experiments (listed in the text). The results from any one of these experiments is likely to cause significant theory updating. These experiments are not complicated; they just need some resources. Compared to what can be learned from these experiments, the resources required are negligible. Unfortunately, I do not currently have the means to perform them.

Nevertheless, I have written six standardized papers on the theory of relativity, astrophysics, and cosmology based on a large amount of scientific data. However, since the complex academic scientific paper is not suitable for general readers, I have used plain language to put the basic content of those papers into this popular science book. Those already published dissertations are in the appendix attached to the back of the book, for readers who have interest in reading further.

If you have the patience to read through this book and then compare it to the attached papers, or even to go and read my 2015 book 谁有权谈论宇宙 *Who Should Talk about Cosmos*, then you will see a clear description of my thinking, which starts from the philosophy, moves through the mathematical papers, then ends back with the literature. This will be helpful to whoever wants to continue in-depth research in related topics or to write something similar.

The modern cosmology, can be traced back to Mr. Einstein. His incredible theory of relativity, has been leading the human exploration in the vast universe for about hundred years.

Originally I did not want to touch this big topic. But was forced to research on relativity for several years. Especially the most recent period, My thought has been in which combination, division, relative. Until on Thanksgiving Day, I finally made my thinking analysis and classification clear on the relevant issues.

Perhaps you will feel novelty and excited because of the criticism on relativity theory, then please enjoy your reading pleasure! Perhaps you will feel angry on our attitude towards the god of science, then, please come up to criticize this book!

I welcome criticism with good intentions, no matter how harsh or even picky. Science and truth become clearer after correct argument. I would also like to remind readers: this book consists entirely of criticism of existing theories. So the reader needs to seriously think in order to discern right from wrong while reading this book.

This book, perhaps would quietly sink in the sea of information, might cause little waves dash; maybe give a little honor, but is more likely to cause me harm. To hell with it! At least it has brought me a lot of good friends. On the way of life, can leave a few lines of shallow footprints in the world, can be regarded as not wasted my life, that is good enough to make me laugh.

This book was written after a lot of effort, after hundreds of modifications, numerous meditations, and tireless reading, and expanded little by little. Most of my spare time is spent traveling, and sometimes I would fall into contemplation beside the Mayan pyramids, sometimes think in front of the grand marvel of the former Summer Palace of the Hermitage in St. Petersburg. And sometimes, I would walk among the pages, information, and lines of this book; though it is long and winding, it has endless, beautiful scenery.

It is: Rainy night, strong wind knocking computer keys,
        Ten years the book caste by light and dark,
        Continue on even though black hair whitened by time,
        The book becomes a brilliant rainbow, sweets my heart.

Yuxiang Wu, Nov. 25, 2016 at San Francisco Bay Area Green Garden
YuxiangWu@outlook.com

PART I

A 4TH GRADE MATH QUIZ THAT HAS BEEN FOOLING HUMAN A CENTURY[2]

> Man thinks; God laughs!
>
> --Ancient Jewish Proverb

I have been wandering in the universe, carrying around star lights, traveling through frenzied starbursts, reading the beauty of the universe, and also learning the strong thoughts of the philosophers, poets, scientists that are filling the space through the flowing time river from ancient to current. The explorations, confused puzzles, wise flames, the shining of the genius..., all of these let me marvel.

There are pieces of puzzling maze, shielding people's sanity. So people caught in the hard exploration, and occasionally bring a breakthrough pleasure. And I wanted to in this ambitious universe, clean out a few pieces of puzzling clouds, and write down some fresh feeling.

These words of the book are so hard to write and spread to the world. They come around from so far away, perhaps already floating in the universe for many years, only by chance encounter you. This is a new song that is sing so old, overripe, tired of the general topic. Please you are willing to sit down and relax for a moment, and read this ray of the freshly colored sunset knocking on the

---

[2] Please refer to following two published papers. They are translated in English with a popular science style in this book:
- Yuxiang Wu, Johnny Wu, "相对论理论蕴含的时间悖论、空间灾难及三个验证实验设计", 格物 2015 第 5 期, No.5, pp 59 - 68
- 吴裕祥 Johnny Wu, '爱因斯坦的"动钟变慢"悖论及"动尺变短"灾难从何而来？', 格物 2015 第 6 期

Part I A 4th grade math quiz that has been fooling human a century

window of the heart.

I searched on Google and got 20,500,000 items within 0.72 seconds.

The relevant content appears in textbooks of secondary school to university.

Countless civil scientists fought on the related topic, produced countless results. So this should be a solemn topic, but was vulgarized as a ladder climbing toward the peak of mankind.

This book is not the same. So I hope you can keep your patience to read several pages first, and then decide whether to continue. Rather than abandoned it as shabby shoes after seeing the name and topic. This time, this book, will give you a fresh landscape, bringing a joyful surprise.

The topic of this part of the book is about the relativity theory that has been dominating the thought of human universe and the scientific community about one hundred years.

Let us take a look, how Mr. Einstein used a fourth-grade level math problem, lead us through the glorious century. From Einstein's foundation paper built the concept of relativity, we have seen that there is something surprising about it, so that it is doubtful that for a century mankind has been provoking God to laugh at loudly.

Where to start?

I have tried to learn and master the electrodynamics equations, digging in Einstein's dazzling math. As if to cure a towering tree disease, know that it is sick but not know where the disease is, so look at its fruit, check its branches, like many people who question the theory of relativity. But did not realized should pay attention to the roots of the tree. After finally I read through the origin context of Einstein's theory of relativity from his groundwork "On the Electrodynamics of Moving Bodies.", the first impression is: how can it dominate the thought of human science about hundred years? People have to reflection on it.

But people are convinced, most are in worship of the relativity theory. The voice from those who oppose, is not convincing; but those stories from who defense for the relativity, are so surprisingly incredibly ingenious. "Only a few people in the world understand relativity." Famous masters put the king's new clothes on the theory of relativity.

The establishment of the concept of relativity is rough, contrary to the basic scientific rules, full of loopholes. As long as you have the patience, even junior

Part I   A 4th grade math quiz that has been fooling human a century

high school culture, can also make the problem clear. But in the past hundred years people who were against it could not refute down it. The main reason, I will be devoted to it later. Here we give a simple explanation: the opinions against the theory of relativity often are based on the basic model from its application, but this is precisely the most unlikely to be successful.

I searched the relevant information and found that there are numerous studies, but few to care about the original source of establishment of the relativity concept. I found the source and had a look: there are so much flaw, the reasoning is so unscientific. In fact, most people, if have the patience to sit down and carefully analyze the original paper "On the Electrodynamics of Moving Bodies" that based on it Einstein established the concept of relativity, can find problems in it.

"On the Electrodynamics of Moving Bodies" is attached in the appendix of the reference section of the book. Only displayed the first three sections. I have checked out the parts that we need to pay attention to in the complicated narrative using underline. This article discusses these important parts. You can go to the original text to see what we focus on, and whether the outrageous or misunderstanding. If you have time, you may want to take a look at the first several sections §1 and §2. If you have the patience to carefully analyze the various notations, instructions, definitions and the proofing process, you can see the problem. The math used here is simple, that is, the fourth grade school math problems of the boat sailing in still water and flowing water. The key is to consider carefully about the legitimacy of the definitions, the physical meaning of the mathematical expression, etc. They are also important points that few people noticed in so many documents discussing relativity theory. The first two sects have no complicated math, and there is no hard reasoning, but it is the basis of the relativistic concept, and the origin of all complex mathematics in relativistic theories. If the foundation is wrong, the building on which it is built collapses; the source is crooked and the intricate branches are totally unhealthy.

The third section of Einstein's thesis is the concrete application of the concept of relativity established earlier, and the mathematical tools are getting complicated. In this section, Einstein derives the Lorentz transformation formula. Just because Lorenz was earlier than Einstein to get the formula, so named Lorentz transformation. This is an important application of relativity theory. From this third section, we found three application problems; and we also found five problems were generated in the process of establishing and proofing of the relativity concept in the first two sections. They are summarized as follows:

Part I   A 4th grade math quiz that has been fooling human a century

Three major problems were found in applications of Lorentz transformation (Einstein transform) in §3 in Einstein's original paper "On the Electrodynamics of Moving Bodies":

1. Using the Lorentz transformation (Einstein transformation) to do relativistic calculation on three or more different reference systems at the same time, the results obtained will conflict each other.
2. The "moving meter stick is shorter" will cause a disastrous disintegration of the high-speed spacecraft
3. "Moving clocks are slower" cause clocks moving in different directions in a spacecraft indicate different time inconsistency paradox.

Five errors were found in the process of establishing the concept of relativity in §1 and §2:

4. Although the mathematical calculations in the original publication are correct, the physical interpretation of the equations is incorrect.
5. The relativity principle defined in the paper is used to prove that the absoluteness of simultaneity is wrong, i.e. the relativity concept is correct, before the rationality and validity of the relativity principle definition is proved.
6. The theory lacks internal consistency—we can use the relativity principle defined in the paper to prove that the relativistic concept is correct and also can use it to prove that the relativity concept is incorrect.
7. Based on the improper definition of the relativity principle defined in the paper, the paper extends the observation results of the static rod system to the moving rod system, extends the local observations results to systematic results, and gets the wrong conclusion.
8. From the philosophical point of view of the mathematical limit (lim), the relativity principle defined by Einstein equates the fragment results during the event development to the final result; confuses the limit process to the limit result, thus violates the basic mathematical limit philosophical law.

Point 8 is a very important, but more difficult problem.

Of course, there are more problems that can be explored, we only did some initiate work. Hope better work will come up.

We don't have to care about the rest of the "On the Electrodynamics of Moving Bodies" as I did at the beginning researching on relativity. In the rest of

Part I   A 4th grade math quiz that has been fooling human a century

the paper complicated math models are presented. Because if the relativity concept is wrong, its application is just looks funny. The more elegant of the math model, the more mistakes confusing.

Prior to detailed discussion, first there come two interesting quiz problems.

There is no need to answer them right away. After reading this chapter, you know how to answer following problems.

Two interesting quizzes

- Issues associated with flying rabbit, turtle, dog and watches

Please read and consider the following little fable. The story consists of three parts, and each part has a problem. Let's try to answer these questions.

1. A dog can run at the speed of light. He is wearing a watch that is always accurate (i.e., it keeps time and never runs fast or slow). He is running back and forth on an east-west-oriented runway at the University of California at Berkeley.

Problem 1: When will the watches lose the correct time? (It is difficult to answer.)

2. A rabbit can fly at the speed of light. He is wearing the same watch as the dog, and is located at one end of the east-west-oriented runway of Shanghai of China, starting to catch up to a turtle at the other end of the runway, who starts running at the same time. The turtle can run very fast, but of course not as fast as the rabbit.

Problem 2: When will the rabbit's watch lose the correct time? (This is still difficult to answer.)

3. Now **imagine** that the rabbit, turtle and dog start running at the same time, and compare their movements.

Problem 3: When will the watches lose the correct time? (Figured it out yet?)

If you still haven't gotten the answer, keep reading. After you finished reading this chapter, you'll know the answer. (Hint: it involves the theory of relativity.)

- Come back a small interesting spacecraft cabin length quiz

Is it possible to make the cabin of a spacecraft extend and shrink like an accordion?

CHAPTER 1

SPECIAL RELATIVITY QUESTIONED

— Analysis of Einstein's research paper "On the Electrodynamics of Moving Bodies," in which the relativity theory is established

What is the difference between Einstein's relativistic principle and the relativity of our daily life?

We will use the following two graphs to illustrate the normal relativity and Einstein's relativity.

Figure 1.1 is a diagram of the relativity concept that we usually use. In the figure, the stationary person, cyclist, care, airplane, and spaceship all use the same common time, what Einstein called "absolute time." Even if these people carry watches to measure time, the hands of these watches will point to the same moment in time and will not change due to the relative comparison of the stationary system to the moving system. However, the people moving in the different vehicles will not feel that their speeds are the same when they observe other moving objects. The person riding the bike will feel that the people who are driving are moving very fast, and those who sit on the plane will feel that the people who driving are moving very slowly. This is the normal relative feeling for a normal person.

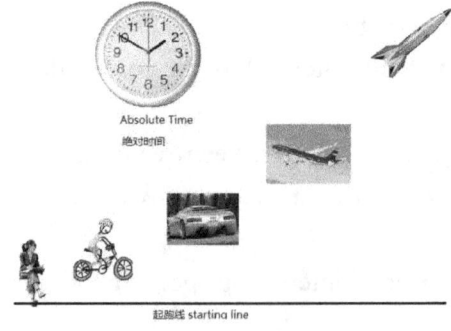

Figure 1.1. All watches on any means of transport follow the same absolute time. Otherwise the rules of time will be meaningless.

Figure 1.2 is Einstein's diagram of the relativity concept. Look the people in the figure. They are sitting, cycling, driving, and flying. All of them will be using the same absolute time as the people in Figure 1.1.

However, in his research paper "On the Dynamics of Electromagnetism" Einstein proved that if they compare themselves to each other, and then their respective watches will point to different times. How did he prove it? We will have a detailed description refer to Einstein's research paper. And his theory of relativity is based on this erroneous ideas, leading to a series of theoretical and application errors.

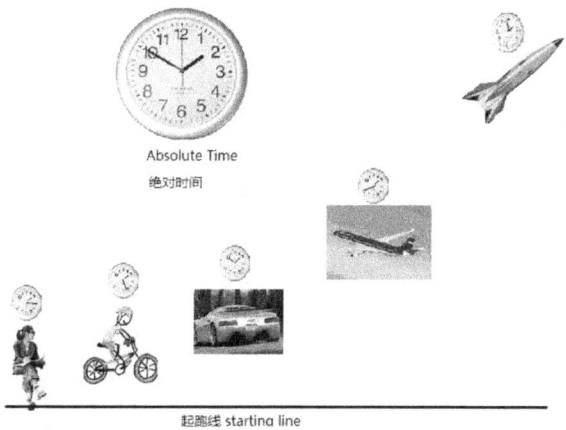

Figure 1.2. But Einstein's theory of relativity argues that the watches each person in the figure has would point to a different time when they move relatively to each other.

Now we just roughly intuitively understand Einstein's relativity concept, and build a preliminary impression.

We can take any pair, or a few different moving objects, in Figure 1.2 to compare their relative movements.

For example, for simplicity, as shown in Figure 1.3, imagines that the sitting person and the person on the spacecraft are separated out to study their relative movement.

First of all, we have noticed, the person is sitting and the person on the spacecraft are exactly the same as in Figure 1.1. Their movement states and movement modes haven't been changed in the slightest.

Imagine attaching the sitting person to a very long, non-deformed, rigid rod, and at both ends of the rod, there is a mirror. A laser light beam is sent from the

Part I A 4th grade math quiz that has been fooling human a century

starting line where the person, sits moves to the rigid rod other end, and immediately reflected back by the mirror at the end of the rod.

Also imagine attaching the spacecraft to a very long, non-deformed, rigid rod, at both ends of which are mirrors. A laser light beam is sent from the starting line, where the person sits, moves to the rigid rod's other end, and is immediately reflected back by the mirror at the end of the rod.

Note that the rod, laser, and spacecraft are completely independent of each other without interference.

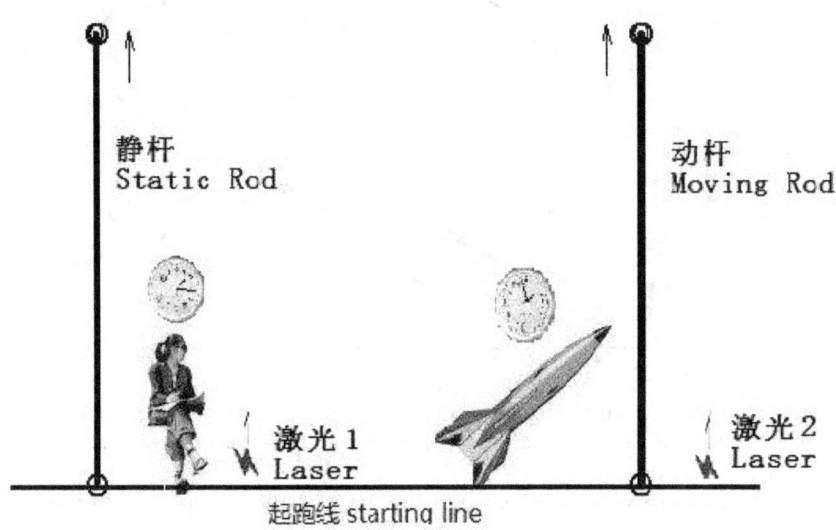

Figure 1.3 the static and dynamic system diagram after imagine adding of rigid rods and lasers

Now, on the starting line of Figure 1.3, there are four objects: the person with a static pole, the laser 1, the spacecraft moving together with an attached rod, and the laser 2. They start moving at the same time and in the same direction.

When the objects are moved to a certain point, such as to when the laser beam reaches the end of the static rod, the state of the system is as shown in Figure 1.4 below.

At this point, the laser 1 reaches the end of the static rod and is reflected back by the mirror at the rod's end; the laser 2 has not yet reached the end of the moving rod, and it needs to continue moving toward the moving rod's end.

24

Laser 1, laser 2, and the moving rod continue to move until the laser 2 catches up with the moving rod's end mirror and is reflected and moves to the spacecraft at the other end of the rod. We can take the above as an observing stage.

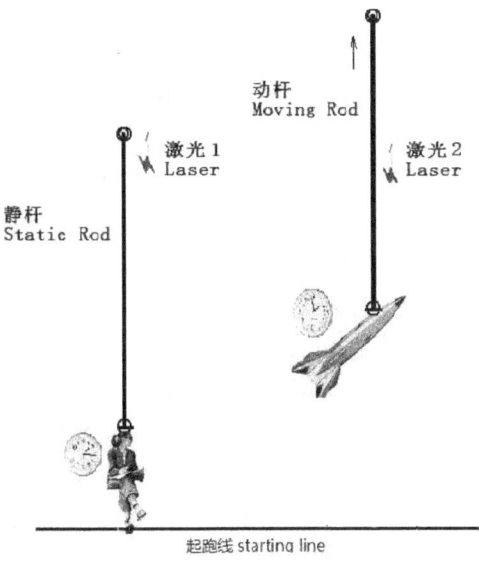

Figure 1.4 the state of the system when the laser moves to the end of the static rod

Question: At the completion of this study stage, will their hands on the two watches, the one on the wrist of the sitting person and the on the spacecraft, point to different moments in time??

The answer of the normal person is definitely something like this: the two watches will indicate the same time. Looking back at Figure 1.1 and Figure 1.2, if the times indicated by the two watches are different, the world has no definite time. All the moving watches will point to different moments in time – this means that all we know about the world, and all of our experiences, are completely different and in conflict!

We also have to note that either in the laboratory or in real life, in Figure 1.1 and 1.2, for the persons sitting, cycling, driving, flying by plane, or flying by spaceship, and the lasers in Figure 1.3 and 1.4, the movements between them are actually completely independent, and it is impossible to tie any two together. The so-called relativity between different moving systems is entirely carried out in the imagination of subjective consciousness! So, in Figure 1.1 and 1.2, in the

Part I   A 4th grade math quiz that has been fooling human a century

imagination, take any two moving objects to make the relative comparison, and you will get basically similar results!

But if the existence of the relative system is only in the imagination, how can the watches possibly indicate different times?

The magic is that Einstein uses simple math (this calculation is as simple as a fourth-grade math problem in which one ship sails in static water and one in moving water) but with a variety of dazzling definitions, terms, assumptions, symbols, etc. In his paper "On the Electrodynamics of Moving Bodies," he made what seems perfect proof: he proved that the time that is indicated by the sitting person's watch is different than the time indicated by the moving watch. And it is this magical difference that has become the most fundamental basis of relativity theory and that has been guiding human science for a hundred years.

In this chapter, we will carefully review Einstein's thesis and reveal how he seems to have successfully done this magical proof.

Einstein used this magical proof to establish his concept of relativity, and then he established the theory of relativity based on this concept.

As the theory was based on an incorrect concept, it cannot be correct.

This chapter will first point out the three major problems in the application of relativistic theory and then examine Einstein's paper "On the Electrodynamics of Moving Bodies" to explore how the concept of Einstein's relativity is wrong and to show where the problem lies.

Relativistic Application Errors Are Caused by Wrong Relativity Theory - Three Fatal Errors in LORENTZ Transformations

Let's take a look at the three fatal relativity application errors in Einstein's relativistic theory, and then analyze Einstein's original paper on the concept of relativity and find out how these error were produced.

To avoid duplication, we first explain the Lorentz transformation theory that has been used since the conception of the theory of relativity. It was not called the "Einstein transformation" because Lorentz proposed it earlier than Einstein. In the Appendix of this book, in Einstein's original paper of special relativity theory of "On the Electrodynamics of Moving Bodies" we can see that Einstein reached the same conclusion without knowing that Lorentz deduced the transformation theory. The coordinate transformation used in Einstein's special theory of relativity, is called the Lorentz transformation. So we know that the

Lorentz transformation is actually an important part of and foundation of Einstein's theory of relativity.

Introduction to the Lorentz Transformation

In 1904, the Lorentz transformation was proposed to explain the Michelson-Morley experiment that was performed in 1887. In that experiment, the movement speed of the Earth relative to the Ethernet reference system couldn't be measured. According to Lorentz's vision, when the observer is moving relative to the ether with a given velocity, the length of the ether (i.e. space media) would contract in the direction of the motion, offsetting different parties differences in the speed of light, so that would explain the Michelson-Morley experiment's zero result.

Einstein later in his groundbreaking theory of relativity "On the Electrodynamics of Moving Bodies", derived the transformation used for two relatively uniform motion of the inertial reference frames (S and S') that is basically the same as Lorentz transformation. So the Lorentz transformation also is Einstein's coordinate transformation in special relativity theory. Suppose the axis system for S is X, Y and Z, the axis system for S' is X', Y' and Z' To keep things simple, let X, Y and Z axes are parallel to the X', Y' and Z' axis. S' system moves along the X axis in the positive direction at a constant velocity v relative to S system. When t = t' = 0, S system and S' system coincide with each other at the origin. The space-time coordinates of the same physical event of S system and S' system is linked by the following relationship which consists of four simple algebra equations. Its math is simple. What important is its physic meaning:

$$\begin{cases} x' = \gamma(x - vt) \\ y' = y \\ z' = z \\ t' = \gamma(t - \frac{vx}{c^2}) \end{cases}$$

In which $\gamma = \dfrac{1}{\sqrt{1-\beta^2}}$, $\beta = \dfrac{v}{c}$, v is the velocity of the S' system relative to S system along the positive direction of the X axis. C is the speed of light in a vacuum.

The problem is that when this simple theory is used to describe the motion of the system, the theory itself contains no adjustable conflicts in its theory and

Part I   A 4th grade math quiz that has been fooling human a century

application, leading to "moving stick shorter" disaster, "moving clock slower" paradox and contradict results from multi relative systems. This is what we will first focus on in this part of the book.

1. THE CONTRADICTORY RESULTS OBTAINED WHILE APPLYING THE LORENTZ TRANSFORMATION TO THE RELATIVITY CALCULATION FOR MULTIPLE REFERENCE SYSTEMS

The coordinate transformation between two inertial reference systems (S and S') may use the following Lorentz or Einstein transformation equations:

$$\begin{cases} x' = \gamma(x - vt) \\ y' = y \\ z' = z \\ t' = \gamma\left(t - \frac{vx}{c^2}\right) \end{cases}$$

It may be generally written as following:

t' = f (t, v, x)
x' = f (x)
y' = y
z' = z

Now add a similar system S". Then, S" and S' transformation is as:

t' = f (t'', v'', x'')
x' = f (x'')
y' = y''
z' = z''

Question: Could the two t's or two x's in the above two equation groups possibly be the same? If they are not the same, which value should be taken? If there are 100, 1000… parallel systems with different speeds added to calculate the t' or x', what is a good result value? Einstein did not limit that only two parallel systems could be used, and it can't be limited in the real world, right?

Let's specifically calculate several examples to get more impression.

Assuming at a ground test station doing a relativity test for five spacecraft , calculate the time dilation due to relativity. We use the most easily searched relativistic calculator on the web to calculate. This calculator is provided free of charge by Casio Computer Co., Ltd. and can be found directly on the web. (http://keisan.casio.com/exec/system/

1224059993). There are many such calculators, provided by universities or research institutes.

In the following figure, the calculator is divided into two parts, the upper part of the horizontal line is the parameter input part, the lower part is the result output part, and the lowest part is the relativistic formula which is used in the calculation. The formula is got according to the Lorentz transformation.

Figure 1.4-1 Time Dilation Calculator

In this calculator, for the sake of simplicity, we fixed the movement time of the object for 1 second and entered a different relative velocities. The observer then obtained a 1-second motion study period, calculated according to the Einstein relativistic theory of time dilation formula, and obtained different time dilation results. Of course, our spacecraft can not have those speeds of movement, here are only used to illustrate the problem.

The result of this calculator can be accurate to 50 decimal digits. We rounding to three. For ease of comparison, the order of the numbers in the

Part I A 4th grade math quiz that has been fooling human a century

table has been rearranged. The formula is generic to any moving objects. We are here imagine them as the five spaceships. We summarize our calculations in following table:

Elapsed time at a body T_0	1	1	1	1	1
Elapsed time at observer T	1.061	1.155	1.342	1.812	3.945
Velocity ratio to light v/c	1/3	1/2	2/3	63%	96%
Relative velocity v	10	15	20	25	29

This is something that thousands of students or professors have calculated, nothing surprising, and no problem.

Now, a little further, please think carefully about my question:

If the observer at the same time take relativity observes on the five spacecrafts that each has different speed, the observer's clock pointer should point to where?

This is a question no way to answer.

Actually, this is the fundamental problem of Einstein's theory of relativity! When he was establishing the concept of relativity, Einstein believed and proved that if the observer was relative to the first spacecraft, his clock would be slower by 0.061 seconds; while the observer was relative to the second spacecraft, his clock would be slower by 0.155 seconds ...

But he did not expect, and did not prove, that if the observer is doing relative to all five spaceships at the same time, the observer's clock pointer should point to which of the five relative times.

In fact, he can not answer. Because it is the wrong result derived from the wrong concept.

Did Einstein really said that? Is Einstein's theory of relativity really so ridiculous? Someone would say: had you made a mistake? It is only calculating it. Even calculate according to Newton's law, will get different times. But they are just a relative feeling, not has really changed the time of a clock.

This is what I said that the concept of relativity of ordinary people is fundamentally different from Einstein's relativity! According to Newton's law, we just calculate, get out at different feelings for relative speeds. As mentioned earlier, the people in a plane watches the movement of a car is very slow; a walking person says the movement of the car is very fast. But the time of the watches of these people wearing do not change.

But Einstein's theory of relativity is different, he defined and proved that, while doing relative to each other, the clock pointers of objects with different movements would be pointing to different time moments! He proved in his research paper that do a relative movement observation the time will change!

Do not say that this is ridiculous, in fact, contemporary cosmologists are also so believe!

They not only believe, but also proved with their experiment, that in the case of relative movements, the clock time would be changed! Cosmologists have done a number of experiments to confirm the time dilation, such as people often cited flight atomic clock experimental (Hafeleand Keating, 1972.) Unfortunately, they did not prove that if in this flight atomic clock experiment, using two or more different spacecraft with different speeds to do the experiment, the time of the ground clock would change according to which speed of the spacecraft!

By the way: this relativity, only, but can only, be made in the ideological. Can you tie them together? Is the thought so powerful?

The task in this chapter is to find out the root of Einstein's erroneous ideas by analyzing of Einstein's original research thesis!

The following is using the same reason from the same source to get the formula to calculate the relative speeds. However, this calculation seems not easy for people to get the idea the author want to express. So that there was an important astronomical experts who had not understood that this example of the relative speeds was the same as the above example of time dilation. Thanks to this expert's reminder, I added above time dilation part. The following example of relative speeds follows exactly the same relative principle. But it is not so intuitive as the time dilation example. You can skip reading it, or try to read to see if it is really so difficult to understand..

Below is an example of calculating relative speeds, the principle is exactly the same as the relative time calculation, but it's not so intuitive.

Part I   A 4th grade math quiz that has been fooling human a century

Look at A, B, D three objects in the following system box:

```
                    VA
      VB       --------> A
--------> B
                    VD
      ------------------> D
```

If object A moves at speed VA in the same direction as object B, and object B moves in the same direction as object D at velocity VB.

Then the velocity of object A relative to object VD is calculated by the following formula:

$$VD = \frac{VA + VB}{1 + VA*VB / c^2}$$

These are commonplace. There is no different from the old ways that people have been doing for a many years.

Wonderful is from now on.

The different from convention is, two more objects A1 and A2 are added to above system as show in below box. And all 3 objects A, A1 and A2 are doing same movements with different velocities VA, VA1 and VA2 relative to object D simultaneously. According to the same calculation formula, we can get different relative velocities of object D relative to object A, A1 and A2.

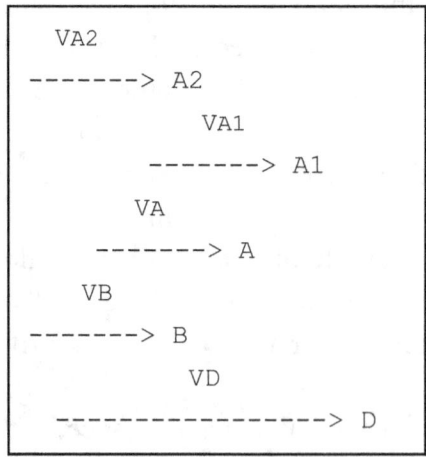

32

When three objects A, A1, A2 are relative to the object D to do relative movement with different velocities, VD at the same time has three values VD-A, VD-A1, VD-A2.

VD should take which value? If a clock is placed on D, which time the clock will point to?

Now we use the specific data to calculate the results. There are a lot of similar materials on the network, readers can check on their own. Moreover, there are many special relativity calculators, just fill in the simple data, you can get the desired results. The following data and results are calculated based on the following calculator:

http://keisan.casio.com/exec/system/1224059837

First determine the speed of the object B: VB = 280000 km / s.

Given the VA, VA1, VA2, we can calculate the corresponding relative compound speed VD as follows:

When VA = 100,000 km / s, VD-A = 289,735 km / s
When VA1 = 200,000 km / s, VD-A1 = 295,733 km / s
When VA2 = 250,000 km / s, VD-A2 = 297,944 km / s

Question: When three objects A, A1, and A2 move at different speeds VA, VA1, VA2 relative to object D at the same time, we calculated that VD has three different values VD-A, VD-A1 and VD-A2. So, what value should VD take? Which velocity should the pointer of Einstein-placed speedometer on D point to?

Similarly if there are 10, 100 ... objects doing relative movements relatively to object D at different velocities, what is the relative velocity value VD of object D? The calculator given above, not only calculates the relative speed, but also calculates the relative time and relative length. In a similar way, it can be calculated that D will have multiple time values at the same time with respect to multiple relative objects, or have multiple different shrink lengths at the same time. So, if Einstein let the observers observe the clock on the object D, the clock pointer will point to which moment?

I know some scientists put X clock in the space shuttle, and Y clock synchronized with X on the ground. After X clock traveled back from the space, the time difference was found between X and Y clocks. This was claimed that it proved the correctness of relativity.

Part I   A 4th grade math quiz that has been fooling human a century

Now let's consider such a scenery: There are three synchronized clocks X, Y and Z. X is in a spaceship, Y is on the ground of the Earth, Z is on the ground of the Moon.

First in our thinking we only pay attention to X, don't consider its comparison to Y or Z. Will X has time error? If you say yes, how to calculate it?

Then in our thinking compare X clock to Y clock, we get X clock time TX_{XY} and the time difference T_{X-Y}. The value of T_{X-Y} is calculated according to Lorentz Transformation formula.

Now in our thinking compare X clock to Z clock, we get X clock time TX_{XZ} and the time difference T_{X-Z}.

The pointer of X clock should point to TX_{XY} or TX_{XZ}?

This is a contradictions without solution. But in fact, all the pointers of the clocks on the objects are pointing to one time point only!

These contradictory overlapping results, only brought about by the illusion of "feeling" of Einstein's error relativity principle. These contradictions arise from the distinctive relativity concept of Einstein shown in Figure 1.2. Where the relative movement will change the time between each other, then a number of relatively moving objects at the same time doing Einstein's relativity calculations, of course, there will be contradictory results.

This contradiction can be verified by experiments similar to those described later, and the reader is encouraged to consider designing and doing their own experiments.

The most incredible Montage is: the relative comparisons of X, Y and Z clocks are happened "**in our thinking!**" All those "relative objects" are only relatively happened inside one's thinking, but scientists take it as a real scientific principle! As just discussed above, if the moving objects A, A1, A2, B, D, are independent from the movement of each other, their speeds will not change because of the movements of other objects; if put some clocks on the rods, the times indicated by the clocks will not be confused. Now, in the mind (note that is in one's thought!) group up them relatively, their speeds thus changed! The clocks on them indicate different times! Their lengths will be changed! What a powerful thought! Is it the thought of God?

As always, certainly someone or experts and professors will come up wonderful ways to explain the above contradictions. But the evidence found in Einstein's original paper makes no excuse for any sophistry! Please be patient.

2. "MOVING METER STICK IS SHORTER" DISASTER

"Moving meter stick is shorter" will have catastrophic consequences for the aerospace navigation, as it basically negates the possibility of humans using high-speed spacecraft.

First, "Moving meter stick is shorter" is systemic change.

According to the Lorentz transformation theory, "Moving meter stick is shorter" is such that in a high-speed flying system with its movement in the X-axis direction will change the length only in X-axis direction, but the lengths in the directions perpendicular to the X-axis remain unchanged. (x' has changed, y' and z' are unchanged.)

Thus there is a relativistic motion catastrophe: if a spaceship is flying at high speed follow X-axis direction, then according to "Moving meter stick is shorter", all components on the spacecraft, including the spacecraft itself, will become shorter along the X-axis direction. This would result that the spacecraft cannot fly and may even disintegrate.

How big is the change on the X-axis?

If the speed of a spaceship is the 90% of the speed of light, the length is calculated in accordance with the Lorentz transformation that a dynamic original foot length after its shrinkage becomes 44% of a foot, or 5.3 inches. In other words, a meter-long rod in the spacecraft moving with 0.9 light velocity will become less than half a meter in length. And the changes only happen in the single direction of X-axis. Thus, significant structural damage to our hypothetical spacecraft is inevitable.

Let us look at some of the results below to get more impression:

Speed of Spaceship	Observed Length	Observed Height	Observed Width
0	200 m	40 m	60 m
10 % of the speed of light	199 m	40 m	60 m
86.5 % of the speed of light	100 m	40 m	60 m

Part I   A 4th grade math quiz that has been fooling human a century

99 % of the speed of light	28 m	40 m	60 m
99.99 % of the speed of light	3 m	40 m	60 m

If the speed of a spaceship is 14% of the speed of light, then the spaceship length along moving direction of the X-axis will be shortened by 1%. Since this is a systemic contraction, regardless of materials used in the spaceship, we have no way to prevent it. It would cause big engine trouble.

Therefore, the transformation formula that has been used for a century in Einstein's relativity systems, its result is permanently blocking our hope of high-speed space travel, blocking our space explore dream!

So, it is essential to find out whether "Moving meter stick is shorter" is correct or wrong.

Experimental design for testing "Moving meter stick is shorter"

First ask contemporary physical gurus a question: Now, do you want Einstein's theory is right or wrong? If correct, then humans no hope to embark on high-speed cosmic adventures; indeed, may never be able to fly out of the solar system or Milky-Way; if wrong, a longstanding pillar of cosmological science will collapse.

We will now design an experiment to verify if the "moving meter stick is shorter" disaster is correct or not. The primary purpose of the test is to determine whether the "moving meter stick is shorter" occurs systematically rather than the rod itself due to the different materials, and thus at high speeds is affected to different degrees.

In this experiment, we take a variety of rods, such as cardboard strips, plastic rods, steel bars ... are on the same spaceship and are vertically placed in 3 different directions to examine their variations in length. As shown in Figure 1.5, the white rods are plastic and the black rods are steel.

One of the three plastic rods is placed in line with the direction of a forward longitudinal direction, i.e. the X-axis; the other two are disposed perpendicular to the direction of advancing the spacecraft Y, and Z directions. Three steel rods also will be placed like the plastic rods.

Let following are lengths of the rods after the test be:

Lrs-1, the length of the steel rod RS-1,

Lrs-2, the length of the steel rod RS-2,
Lrs-3, the length of the steel rod RS-3,
Lrp-1, the length of the plastic rod RP-1,
Lrp-2, the length of the plastic rod RP-2,
Lrp-3, the length of the plastic rod RP-3,

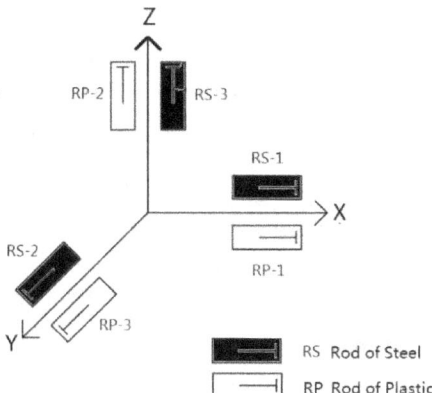

Figure 1.5 Rod test with two rod groups made of different materials for testing "moving meter sticker is shorter" disaster

According to the Lorentz transformation, Since x' != x, but y' = y, z'= z, then there should be:

Lrs-1 = Lrp-1
Lrs-1 ! = Lrs-2; Lrs-1 ! = Lrp-2; Lrs-1 ! = Lrs-3; Lrs-1 ! = Lrp-3;
Lrs-2 = Lrs-3 = Lrp-2 = Lrp-3.

That is, according to all the results of the changes in the lengths of the rods of these two groups participating in the experiment, it can be determined that each inner rod 'length' of the spacecraft system is changed or not in accordance with the relativity theory. If the lengths changed do not meet the relativity theory expected, it should only say like this: the material lengths are affected to change in the high-speed system, people should be further study of the impact of the law on different materials, rather than generally assert "moving meter stick is shorter."

We are sure that the results of this experiment will prove the system length has not changed, but moving objects of different materials within the system perhaps would produce different degrees of impacts by the speed, and therefore the lengths of different materials were possibly changed with varying degrees.

Part I   A 4th grade math quiz that has been fooling human a century

Einstein's "moving meter stick is shorter" theory is not correct. This sentence originally was 'we expect', now is replaced by 'we are sure', because we found from Einstein's original paper of building relativity concept that he made the wrong derivation that resulted in the "moving meter stick is shorter". He used the correct mathematical calculation but the wrong physical concept.

One possible catastrophic consequence of Einstein's "moving meter stick is shorter" mistake is: No matter how spacecraft are manufactured, they cannot work properly at high speeds and may even disintegrate, because all spacecraft components will be deformed as the X-axis shrinks. Therefore, we must do the relevant experiments to resolve this life and death event! If we strive to improve the speed of the spacecraft, but toward the end the spacecraft disintegrated eventually, would not it be funny?

Thus, it would be more in line with the interests of humanity if Einstein's relativity theory is not correct.

Perhaps this says more scientific: in high speed moving system, different materials perhaps would be affected by high-speed movement may have varying different degrees of length changes, plastic rod length change is such, and steel rod length change is such ... this is the need to continue to study.

3. "MOVING CLOCKS ARE SLOWER" PARADOX

In Lorentz transformation " $t' \neq t$ " tells us that, in relative motion, the whole system time will change!

A system of "time" change is not a clock time indication of change can be proved. We have to prove everything in this system, including its time course changes, and the life of everything is changed in the same proportion.

For example, if in a magical spacecraft flying at the speed of light, because of the system speed, the system time is slower by half than outside the system. Then the clocks will go slower by half, the aging process of the pilots would be halved, even the fixtures and equipment would wear out half slower ... as long as there is one thing whose life is not half-time as the time flow rate and should be different, we cannot say that the system time has changed. The Lorentz transformation tells us to change the system time (t' has changed) rather than the clock time only.

But this is actually almost impossible to prove. Who can prove that in the system "all" things have slowed down with the clock and are changing at the same rate? Those in the spacecraft who might have heart disease may be affected by the high-speed movement, possible untimely death, but are their brothers and sisters having same disease on earth still alive? Should we just allow people who are not sick to enjoy this age-slowing treatment?

The problem is not as dire as "moving meter stick is shorter" disaster, but it also needs further research.

Experimental design for testing "Moving clocks slower"

We will now design an experiment to verify if "moving clocks slower" is correct. The primary purpose of the test is to determine whether the "moving clocks slower" occurs systematically rather than by different moving directions, and thus affects different time frames at high speeds.

In this experiment, we take a variety of clocks that are doing axial reciprocating motion, such as mechanical clock, photon clock, laser clock…, put them on a spaceship, and place them in different directions to examine their time variations.

As shown in Figure 1.6, there are two different sets of clocks: a group of mechanical clocks that count time by reciprocal movement, and a group of photon clocks that also count time by doing reciprocal movement. In this way, we not only examine the "moving clocks slower" theory is correct or not, but we can also understand high speed motion systems have what kind of different effects on movements of mechanical or optical clocks system manufactured from different materials. Of course, the ground would use the same kind of clocks to do the comparison, which is not explained.

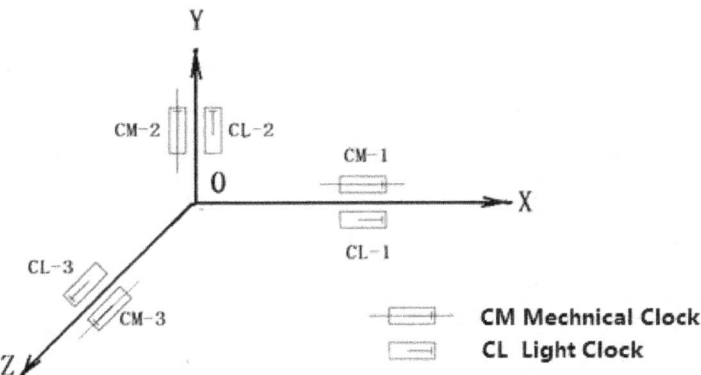

Figure 1.6, two group clocks to test "moving clocks slower" paradox that arises in the Lorenz transformation

Part I A 4th grade math quiz that has been fooling human a century

In the spacecraft one of the three mechanical clocks is placed so that its time movement is in line with the direction of a forward longitudinal direction, i.e. the X-axis. The other two are placed perpendicular to the X direction of advancing spacecraft, at Y, and Z directions. Three photon clocks also will be placed same like those mechanical clocks in different planes and perpendicular directions.

The following are the times of the clocks after the test:

Tcm-1 is the time of the mechanical clock CM-1,
Tcm-2 is the time of the mechanical clock CM-2,
Tcm-3 is the time of the mechanical clock CM-3,
Tcp-1 is the time of the photon clock CP-1,
Tcp-2 is the time of the photon clock CP-2,
Tcp-3 is the time of the photon clock CP-3,

According to the Lorentz transformation, Since T' != T, we should have:

Tcm-1 = Tcm-2 = Tcm-3 = Tcl-1 = Tcl-2 = Tcl-3

However, according to the Lorentz transformation, since x' != x, but y' = y, z'= z, and the computation time is determined by the internal clock that is making a linear reciprocating motion, in the X direction the clock influenced by the spacecraft motion the time change will happen, but in Y, Z directions there should no moving effect and should still be like the original, and indicating times not change should occur. It should be:

Tcm-1 != Tcm-2;   Tcp-1 != Tcp-2;
Tcm-1 != Tcm-3;   Tcp-1 != Tcp-3;
Tcm-1 = Tcp-1;
Tcm-2 = Tcm-3 = Tcp-2 = Tcp-3.

That is to say, if the times indicated by all clocks in both groups participating in the experiment are consistent with those above equations and inequalities, then we can say that high-speed motion within the system of mechanical movement or motion of photons made system is affected. We can further study the effects of related phenomena.

The same principle can be applied to another set of photon clocks. In accordance with the principle of the system time changing, all the photon clocks should indicate the same rate of deceleration. But according to the Lorentz transformation of "moving clocks slower", the time indicating the X-axis direction should be different from those of Y-axis and Z-axis direction. The time indicated in Y-axis and Z-axis direction should be the same.

I believe this result will prove that the system time has not changed, but the states of motion of different objects within the system are affected, thus different clock movements may occur in different variations.

According to the Lorentz transformation, only in the direction of the system moving will the indicated time of the clock slow down, because only in this direction has X' changed. But according to the theory of relativity, all clocks in the system should slow down at the same rate, because t' changed. Thus there is a paradox of relativity in "moving clocks slower": If all of the times indicated by all clocks involved in the experiment changes have taken place in the same consistent, preliminary proof system of "moving clocks slower," the Lorentz transformation theory is wrong; if the Lorentz transformation is right, then the systematic "moving clocks slower" conclusion will be wrong. Test results may prove both wrong, which would result in the emergence of a new law of motion.

This goes back to the problem of how to analyze the systems. Should we break down the complex systems to study them, or combine them together to do research?

Perhaps it is more accurate to say that in the high-speed movement of the system, mechanical clocks or photon clocks affected by velocity will be subject to a different variation. Its law of motion is such …; the human aging process will be delayed in the high speed system. Its law of motion is such … Again, this will need further study.

These are the ideas before 2015. Since it had been written down, I do not want to change it, but rather let it as the historical footprints of my long exploring in the fog for a decade.

Now (November 25, 2016) I realized: the original of these experiments, their principle is right, everything is right, only the results would be very embarrassing! Such experiments, all the experiments on relativity, whether did on the plane or in the spacecraft, are impossible to reach the desired results! Why? Later we will devote to explain it.

So relativity theory is almost impossible to be proved by physical experiment.

Where do "Moving Clocks are Slower" Paradox and "Moving Meter Stick is Shorter" Disaster Come From?

We discussed in detail the "moving clocks are slower" paradox as well as the "moving meter stick is shorter" disaster embedded in the Lorentz transformation

Part I   A 4th grade math quiz that has been fooling human a century of Einstein's relativity theory.

But why is there a "moving clocks are slower" paradox at all? Why is there a "moving meter stick is shorter" disaster? Is it natural law or only a theoretical error? Or even just an illusion of a great man?

Keeping these issues in mind, we once again read the foundation paper of Einstein's relativity theory: "On the Electrodynamics of Moving Bodies." After repeated deductive reasoning, we determined that the problem lies in the section §2 "On the Relativity of Lengths and Times" of the "KINEMATICAL PART" of Einstein's research paper. And in §3, the Lorenz-Einstein transformation was derived from the concept in §1 and §2, of course, inherits the errors of the previous sections, so it is not surprising that the applications of the Lorentz-Einstein transformation full with errors.

For the sake of the reader's convenience, in the book we have attached to the appendix the first three sections in "I.KINEMATICAL PART" of Einstein's "On the Electrodynamics of Moving Bodies" research paper. There are total of five sections in "I. KINEMATICAL PART" of the paper, the 3 we attached are:

§ 1. Definition of Simultaneity

§ 2. On the Relativity of Lengths and Times

§ 3. Theory of the Transformation of Co-ordinates and Times from a Stationary System to another System in Uniform Motion of Translation Relatively to the Former

We will mainly discuss sections §1 and §2, because these two sections are the basis of the established theory of relativity. As for the other sections, §3 is substantially a repeated derivation of the Lorentz transformation, and contains further application of §1 and §2, and from them, the theory of relativity was obtained. Sections §4 and § 5 are based on the conclusions of the previous sections and are simple expansions using partial differential equations.

One serious problem is that if the simple mathematical reasoning in §1 and §2 is based on doubtful assumptions, then the complicated mathematical reasoning follows in §3, §4, and §5 will not be reasonably possible. Regardless of whether simple or fantastical mathematical models are applied to the theory later, they will all become a pile of useless, wrong models because they are all established on the wrong basis.

Einstein made several fatal mistakes in his foundation research paper for building his relativity theory, which caused his concept of relativity to be incorrect. These errors are relatively simple, intuitive, and easy to find. But why has a century passed without these errors being made public? This is a worthy subject for scientific workers to ponder and sum up!

We have listed the three mistakes associated with relativistic applications of the Lorentz-Einstein transform: the contradictions of the results of the calculation; "moving meter stick is shorter" caused disaster; and in the "moving clocks are slower" of the time paradox. All of these can be verified through experiments. Of course, these experiments are very difficult, the reason is discussed in later chapter in detail.

The following discussion is to find out why these three errors appear in the application of relativity from Einstein's original paper "On the Electrodynamics of Moving Bodies."

Before starting our discussion, let's start with a simple closed system. A closer look at this system model will help to better understand the nature of the dizzying model after having a combination of various conditions and definitions.

And then point by point analysis the five major problems generated during the process of establishment of relativity concept that causes the disaster and paradox in its application.

A Basic but Important Model of the Underlying System We Constructed that is Used for Comparison Purposes

The system consists of two rods, two lights moving at the speed of light C, and a few clocks that are always correct and do not interfere with each other. Remove the person connected to the person in Figure 1.2; also remove the rod attached with spacecraft, and assume that this rod still moves at the speed of the spacecraft; thus forming the system we want to study as shown in the following figure.

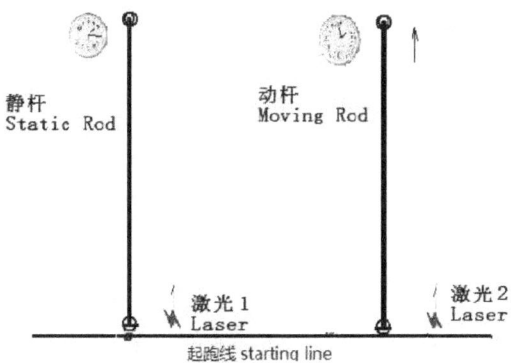

Part I   A 4th grade math quiz that has been fooling human a century

Figure 1.7 The system modified from that of Figure 1.2

We also stipulate that the all objects and all the conditions of the system are the same set as Einstein's rod systems in his paper. **The objects are completely independent of each other**. We put these separated independent objects in the same closed system to study.

When we use the four-dimensional object world line graph to describ the rod system in Figure 1.7, in order to be consistent with the tradition, we rotated Figure 1.2 clockwise of 90 degrees, and added the time axis T, the space axis X, and the corresponding correlation symbols, and got the following Figure 1.8. The four-dimensional object world line graph will be defined and discussed in later chapter.

Simply to say in this section, we use two non-interference rods, two beams of light to do some research. The description in following is complex, because it is to make it clear by using the mathematical calculation, so the relevant names, symbols are essential. If you do not want to explore the details, then know that the following is to put unrelated two rods and two beams of light together for comparison is enough. Then you can jump directly to the next section.

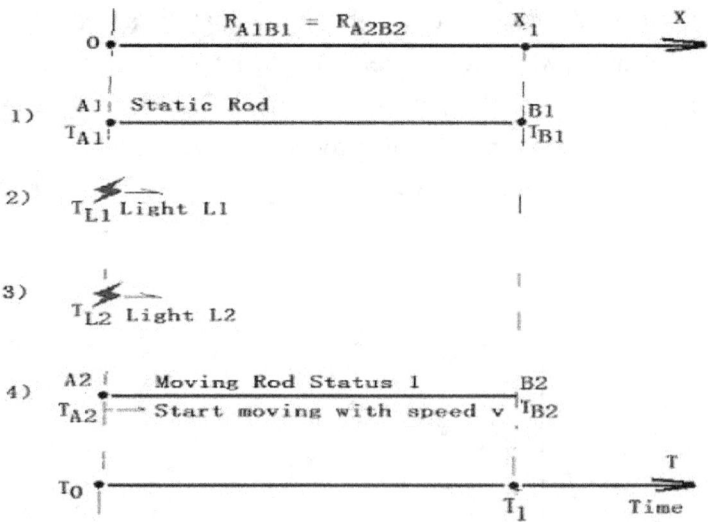

Figure 1.8 Schematic diagram of the initial system status 1 of the basic model we constructed

In Figure 1.8 there are following objects:

One of the two rods is static, the other is doing a uniform linear motion.

One of the two beams of light running back and forth between the ends of the stationary rod without stopping; another beam is moving between the moving rod ends.

These moving objects begin to move at the same time.

At each end of the rod, place a never-erratic clock.

These clocks are synchronized at the beginning of the movement.

When the light moving above the static rod reaches the other end of the rod, are the clocks on the two rods indicating the same time?

When the light moving above the moving rod reaches the other end of the rod, are the clocks on the two rods indicating the same time?

Question: When will these synchronized clocks begin to indicate different times?

So many pages of the following narrative, in fact, is such a point content.

Figure 1.8 shows the initial state of the system. The top is an X-axis for recording the positions, followed by the four objects, and the bottom is the T-time axis for time records.

The four objects being studied are as follows:

1) Static rod A1-B1 is stationary, and its two ends are called A1 and B1. The length of the static rod is denoted as R_{A1B1}. There is a clock at each end of the rod. The time of the clocks are T_{A1} and T_{B1}.

2) The light L1 moves from A1 to B1 and returns immediately after reaching the B1 terminal. The time indicated by the clock of the laser light L1 is T_{L1}.

3) The light L2 is moving in the direction from A2 to B2 in the direction in which x increases along the X-axis, and the time indicated by its clock is T_{L2}.

4) The moving rod A2-B2 starts moving at a speed v along the X-axis from 0 to X. The moving rod ends are called A2 and B2 respectively. The length of the moving rod is denoted as R_{A2B2}. Both ends of the moving rod have a clock, which indicate the time for the T_{A2} and T_{B2}.

At the start time shown in the figure, the lights L1 and L2 and the moving rod R2 are starting to move in the X-axis direction at time T_0. All pointers of the clocks are pointing to T_0, that is, $T_{A1} = T_{B1} = T_{A2} = T_{B2} = T_{L1} = T_{L2} = T_0$.

The use of objects and related symbols in the above section is a little bit too much, so we made a table to summarize below:

Table 1.1 Summarize of rod system initial status 1

Part I   A 4th grade math quiz that has been fooling human a century

Object	Name	Name of End	Name of Clock Time	T-Axis	Rod Length
X-axis	X X_i	Origin Point 0			
Rod	Stationary Rod R1	Rod Start End A1	T_{A1}	T_0	R_{A1B1}
		Rod End B1	T_{B1}	T_0	
	Moving Rod R2	Rod Start End A2	T_{A2}	T_0	R_{A2B2}
		Rod End B2	T_{B2}	T_0	
Light	L1	Starting Point A1	T_{L1}	T_0	
	L2	Starting Point A2	T_{L2}	T_0	
Time Axis T-axis	T_i	Starting Time T_0	T_0	T_0	

Now take L1 arrived B1 event as next system status 2. It is shown in Figure 1.9.

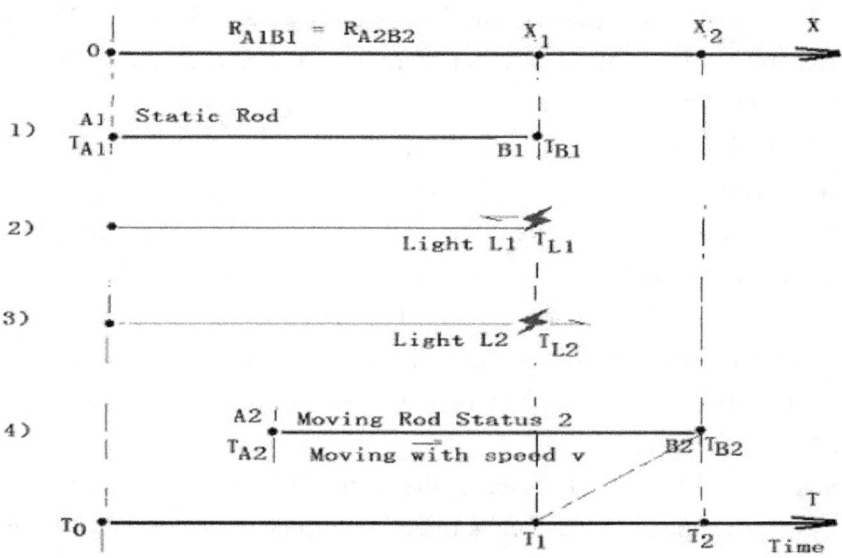

Figure 1.9 Schematic diagram of the rod system status 2 of the basic model that L1 moved to the rod end B1.

At this time, both L1 and L2 are reaching the rod end B1, that is, the position X_1 on the X-axis. But the moving rod ends A2 and B2 are not in the same place;

46

the rod moved forward a distance in which the moving rod end B2 reaches the position indicated by X_2. We've got the following summarizing table 1.2.:

Since the time we recorded is the one at which L1 and L2 reach the rod end B1, T_{B2} remains the same as the time indicated by the other clocks, with $T_{A1} = T_{B1} = T_{A2} = T_{B2} = T_{L1} = T_{L2} = T_1$. That is, all the clocks are still synchronized.

Where the positions of the moving rod ends A2 and B2 are in the table? Please come up with your own thinking.

Table 1.2 Summarize of the rod system for status 2

Object	Name	End Name	Name of Clock Time	X-Axis	T-Axis	Rod Length
X-Axis	X_i	Point 0				
Rod	Static Rod	Rod Start End A1	T_{A1}	0	T_1	R_{A1B1}
		Rod End B1	T_{B1}	X_1	T_1	
	Moving	Rod Start End A2	T_{A2}	?	T_1	R_{A2B2}
		Rod End B2	T_{B2}	?	T_1	
Light	L1	Start A1	T_{L1}	X_1	T_1	
	L2	Start A2	T_{L2}	X_1	T_1	
Time T-Axis	T_i	Start T_0	T_1		T_1	

However, we have to pay special attention to the location of B2 and T_{B2}. At this time, the location of B2 is completely different from the positions of B1. Clock time on laser beams T_{L1} and T_{L2}, which are all at X_1 of the X-axis. But the clocks on moving rod end B2 already moved from X_1 to X_2, T_{B2} is the clock time at X_2, but the time it points to is T_1, as shown in the dashed dotted line connecting T_{B2} and T_1 in above figure.

Now let's play with our imagination. If lights L1 and L2 and the rod continue moving as described above, so that L2 and rod end B2 are moved to the position X_3, is there any clock time pointed to by the clocks that is different? Would there be any time difference between the clocks? If another rod is moved to the end of the equivalent of two rod lengths X_4, or the equivalent of 100, 1000, or 10000 rod lengths, would any clock indicate different time? This is very important. Let's check them one by one.

The clocks on the stationary rod are like the clocks in our house, and they do

Part I A 4th grade math quiz that has been fooling human a century

not indicate the wrong time until they are damaged.

Light L1 moves back and forth between the ends of the stationary rod, just as Michael Morley did in experimenting for checking the speed of light, affirming that the clock time it indicated would not change.

Light L2 moves straight forward and cannot imagine any change in the clock it indicates under the definition of constant light velocity. Of course, if L2 makes the same back-and-forth movement as L1, or goes back and forth for a while and then in a straight line for a while, the time the clock indicates will not be confused.

For the clocks on the ends of the moving rod, we can take the rod to be a boat, aircraft, or any means of transportation traveling around the world. The clocks on the vehicle, in accordance with our knowledge of life, will not be confused. This is the point easy to confuse. Then we might think that the clocks are mechanical clocks that make axial movements or move in a plane (such as a pendulum clock that simply oscillates in a plane). Let the direction of movement of the clocks be perpendicular to the direction of movement or the plane of motion of the moving rod. According to the laws of kinematics, the movement of the mechanical clock will not be affected by the movement of the moving rod carrying it. Since the movement of the mechanical clock is unaffected, these clocks will of course always indicate the correct times.

We can also imagine the moving rod as starting at point A2 and ending at point at B2 of a runway, and its length is R_{A2B2}. At point B2, there is a turtle that runs as fast as the rod. Let L2 become a flying rabbit, which starts flying at the same time as the turtle from A1. The advantage of this is that we do not need to consider such a complex situation like that the length of the rod in motion may change but still can get the same conclusion.

In short, respectively, regardless of how long the continuous movement lasts and how far the distance moved is, whether the motion is linear, or back and forth movement, or curved, these clocks will not be confused and will not point to the wrong time.

Now there is a great scientist who, in his thinking, puts those clocks in different groups. These clocks run correctly and synchronize all the time while they are not grouped in his thinking. But if the scientist compares the times indicated by those clocks in different groups, or adds some observers who will not have any effect on the objects or clocks. After a short period of time, the clocks will indicate different times! These clocks that never make mistakes, then, would be confused by combining them in one's **THOUGHTS**!

Note, ah, that it is the combination of one's THOUGHT and observation as a bystander. What a wonderful idea: magical observations!

Let's take a look at how this century-long miracle of relative confusion happens.

Prior to this, we need to refer to the table above and keep in mind the names of the clocks and the times they indicate, and compare them in the next section. In particular, we note the light beam times T_{L1} and T_{L2} do not exist in the next section. Please think about why we have to add these two time symbols. And why should the static rod clock and the moving rod clock time use symbols such as T_{A1}, T_{B1}, T_{A2}, and T_{B2} rather than general t_A, t_B, t'_A?

The Basic Model of Relativity Theory in Einstein's "On Electrodynamics of Moving Bodies"

The following briefly quote about what we want to discuss from Einstein's first paper built up the relativity theory: "On the Electrodynamics of Moving Bodies". Please pay special attention to boldface. Italics are the originals in Einstein's paper. They correspond to the underlined part of the appendix.

In "§1 Definition of Simultaneity", it says, **in a stationary system**:

*We have not defined a common "time" for A and B, for the latter cannot be defined at all unless we establish **by definition** that the "time" required by light to travel from A to B equals the "time" it requires to travel from B to A. Let a ray of light start at the "A time" t_A from A towards B, let it at the "B time" t_B be reflected at B in the direction of A, and arrive again at A at the "A time" t'_A.*

In accordance with definition the two clocks synchronize if

$$t_B - t_A = t'_A - t_B \qquad (1)$$

"§2 On the Relativity of Lengths and Times"

*The following reflexions are **based on the principle of relativity** and on the principle of the constancy of the velocity of light. These two principles we define as follows :* (author of this book note: The relativity principle was defined at the outset of §2, but the rationality of the definition was not demonstrated or justified.)

*1. **The laws by which the states of physical systems undergo change are not affected, whether these changes of state be referred to the one or the other of two systems of co-ordinates in uniform translatory motion.***

Part I A 4th grade math quiz that has been fooling human a century

2. *Any ray of light moves in the "stationary" system of co-ordinates with the determined velocity c, whether the ray be emitted by a stationary or by a moving body. Hence*

$$\text{velocity} = \frac{\text{light path}}{\text{time interval}}$$

where time interval is to be taken in the sense of the definition in § 1.

...

We imagine further that at the two ends A and B of the rod, clocks are placed which synchronize with the clocks of the stationary system, that is to say that their indications correspond at any instant to the "time of the stationary system" at the places where they happen to be. These clocks are therefore "synchronous in the stationary system.

We imagine further that with each clock there is a moving observer, and that these observers apply to both clocks the criterion established in §1 for the synchronization of two clocks. Let a ray of light depart from A at the time t_A, let it be reflected at B at the time t_B, and reach A again at the time t'_A. Taking into consideration the principle of the constancy of the velocity of light we find that

$t_B - t_A = R_{AB} / (c - v)$ (2)

$t'_A - t_B = R_{AB} / (c + v)$ (2.1) (this book author added formula numbers)

Where R_{AB} denotes the length of the moving rod—measured in the stationary **system observers moving with the moving rod would thus find that the two clocks were not synchronous, while observers in the stationary system would declare the clocks to be synchronous.**

So we see *that we cannot attach any* **absolute** *signification to the concept of simultaneity, but that two events which, viewed from a system of co-ordinates, are simultaneous, can no longer be looked upon as simultaneous events when envisaged from a system which is in motion relatively to that system.*

Questions Raised in Einstein's Research Paper, and the Questions of those Questions

Using the points below, Einstein denies the absoluteness of simultaneity and establishes the correctness of relativity:

1) In the static rod system, t*he two clocks synchronize if $t_B - t_A = t'_A - t_B$ (1).* **based on** the **self-defined** *principle of relativity and on the principle of the constancy of the velocity of light,* **extending (1)** *from the static rod system to the*

moving rod system, there should be (2) = (2.1). Now (2) != (2.1), **so** the absolute nature of simultaneity does not exist, and relativity is correct.

2) Because of (2) != (2.1), ***observers moving with the moving rod would thus find that the two clocks were not synchronous, while observers in the stationary system would declare the clocks to be synchronous.*** *So we see that we cannot attach any absolute signification to the concept of simultaneity, but that two events which, viewed from a system of co-ordinates, are simultaneous, can no longer be looked upon as simultaneous events when envisaged from a system which is in motion relatively to that system.*

We already know that when the moving rod and the light, and the static rod and the light, are separated into independent systems. No matter how long time runs, there will be no problem of clock disorder, and simultaneity always holds. So why, after Einstein, **in his thinking**, groups these objects and compares the groups, there is the problem that simultaneity is not established, and the times indicated by the clocks are different? This idea of the sub-group has so much power. Where is the secret or the paradoxical point?

Using another analysis method

In above "Conclusion," "would thus" and "so" are not convincing. Einstein "cannot attach any absolute signification to the concept of simultaneity" derivation from formula (2) and (2.1). It is not convincing, not fully concluded, and we cannot see why, because he showed (1), (2) and (2.1), then got the conclusions of "would thus" and "so." "Observers moving with the moving rod would thus find that the two clocks were not synchronous, while observers in the stationary system would declare the clocks to be synchronous." We cannot understand why, in a simple comparison with the static system and dynamic system, the two system clocks are not synchronized. Do one or two clocks have a problem? Or do the clocks have no problem, but the observer had an illusion?

When I began to write the second edition of my book "Who Should Talk about Cosmos", I didn't have any intention to touch relativity theory that is such a huge monster, although I don't like many of its ideas (such as space conversion, time traveling, etc.) I was just about commenting on "time traveling" - "If I can run faster than light, I can go back to kill my father before I was born" - felt it was incredible stupidity. Why need be faster than light? Because history hides in the light? I'll not stop questioning this emperor's new clothes because several famous

Part I A 4th grade math quiz that has been fooling human a century

guys said "only a few people in the world understand the theory of relativity." Therefore, I used a funny tone writing down the title of the fifth chapter "Relativistic Literature." Yes, it is a literature rather than a scientific theory.

In fact, whether Einstein said the words of the killing father or not, I do not know. But the "science" principle of the father-killing story should be derived from the theory of relativity. For those Internet people who like the novelty, this is simply too much in line with their taste. So the relativity theory that some people always questioning, with the time traveling stories, excited by Hollywood movies, has been in popularity in recent years.

Because of made fun on Einstein, first of all, "Who Should Talk about Cosmos" could not be published in China, and then the second edition in the Yangtze River Publishing House was stopped by its new president, even the publishing printing edition was completed. It was happened in 2016.

I am a stubborn person, will be regardless of costs to go on for the hobby, and for respectful.

Fortunately, writing a book only need a personal space.

Information required are online.

So I began 2006-2016 long journey of the soul, comprehensive reflect on second edition of the manuscript "Who Should Talk about Cosmos".

But the trouble with relativity theorem never ends.

September 2015, I took the results of a decade of reflection - the book "Debate of light and dark" - to seek publication. It was refused by many so well-known publishing houses because of the problem of relativity. A feedback message said that the answer to Olbers' paradox is convincing, but the critique of the relativistic part is not mathematical enough. And the book can't be published!

Well, from last October to the present, I have been forced to focus on the theory of relativity. Until today, 2016 Thanksgiving, finally get the results satisfying myself.

Reader friends, think about that I am full of displeasure, year after months to be studied relativity, how much tears in my face!

Of course, I also got some interesting results, made myself happy, and the great people of the publishers can have nothing to say!

Scientific publishing should be always about science.

After a big nagging, let's go back to our discussion.

In the following proof, we will use a step-by-step analytic method different from that of Einstein. We will not use the mathematical superposition formula, directly applying it to the velocities, but will instead analyze the moving rod step by step, in each step comparing the times and movements in the two rod systems to find out exactly what time instance observers would find those clocks to show different times. While this makes the analytic process to achieve results and mathematical derivation more complex than what Einstein did, the final conclusion proved that in the Einstein model, the clocks in static and dynamic systems had always been synchronized, but they are at particular times in the different locations given by Einstein by his calculation. The final conclusion proved that the clocks in the static rod system and moving rod system have always been synchronized, but they are at particular times, in the different locations, not as imagined by Einstein. The fundamental reason that caused the error is that, while Einstein divided the system into static and dynamic, in his analysis, he did not consider the physical meaning of the moving rod system, but simply shielded off the fact that the rod location changed after moving by superposition velocity v. Even the basic fact that the end position of moving rods is in flux is not expressed in his model.

We need to know that the meanings and values of "t_A, t_B, t'_A" in §1 static rod system are different from those in §2 of moving rod system, although they have the same symbols. The values of "t_A, t_B, t'_A" are changing with the movement and location changing. The values in the moving rod system are already completely different from the values of the stationary rod system, without the slightest reason to make comparison. But Einstein considered them equivalent in his system analysis. Einstein used a simple elementary school mathematical equation (2) to mask the physical nature of the moving rod system.

Because, according to analysis results of Einstein's model, "relativity" is actually not true, following his "relativity principle" to deduce "moving clocks are slower" leading to paradoxes, is normal. This is the typical example of a wrong model extended out from the same erroneous results.

Before we analyze the details of the moving rod system, let's look at a simple arithmetic problem of rowing in flowing water that fourth-grade students often encounter. Of course, this is not part of our proof, but you can look at it from another angle about the concepts we are discussing.

A boat leaves A at time T_A, heading upstream to arrive at B at time T_B, and then returning immediately downstream to arrive at A at time T'_A. The speed of the boat is C, and the flow velocity is v. If the distance between A and B is named

53

Part I A 4th grade math quiz that has been fooling human a century

R_{AB}, then we have the following relationship (3) and (3.1). We'll focus on (3):

$$T_B - T_A = \frac{R_{AB}}{C-v} \qquad (3)$$

$$T'_A - T_B = \frac{R_{AB}}{C+v} \qquad (3.1)$$

We can easily see that the formation of the equation (3) and (3.1) obtained from the primary school math is exactly similar to Einstein's equation (2) and (2.1). The basic shape is identical; the difference is the velocity. Instead of the moving rod speed, here is the water flowing speed, and the destinations of the boat back and forth are fixed. Can an arithmetic formula from elementary school possibly explain the concept of relativity and the absence of simultaneity? Would the time indicated by the clocks on the ship be different from the time indicated by the clocks on the shore?

Obviously, (3), (3.1), and (2), (2.1) are not the same. Is it likely that the upstream and downstream journey take the same amount of time? Can the upstream boat travel the same distance the downstream boat did using the same time? (2) != (2.1) is an obviously result, how come need using relativity principle to prove that since they are not equal, so the absoluteness of simultaneity is wrong – implies that the relative concept is right!

Is it not a bit unreasonable?

Now we combine the above excerpts of the original Einstein paper to analyze the five specific issues in the paper we mentioned above.

4. THE ERROR IN EINSTEIN'S PAPER RESULTS FROM THE FACT THAT THE MATHEMATICAL CALCULATION IS CORRECT BUT THE PHYSICAL

Now, from the physical point of view with a mathematical analysis, we focus on the rod systems in §2.

In the following analysis, the rod moves with speed v along the X-axis. With D_{SAME} being the distance the light traveled while it starts at the rod end A, goes the same direction as the moving rod, and start at the same time as rod starts, from moving rod end A2 to rod end B2 with speed C along the X-axis. D_{REV} is the distance the light traveled in the opposite moving direction of the rod.

In the equations (2) and (2.1), the distance D_{SAME} and D_{REV} are the same as the rod length R_{AB},

We know from elemental primary school algorithm that D_{SAME} and D_{REV} are naturally not the same.

D_{SAME} is far bigger than D_{REV}. D_{SAME} is much longer than the rod length R_{A1B1}. This value of the distance traveled is determined by the velocity v of the moving rod. Suppose $v = \frac{C}{2}$; then the entire distance the light traveled from A2 to B2 will be $2R_{A1B1}$.

D_{REV} is much shorter than the rod length R_{A1B1}. This value of the distance traveled is also determined by the velocity v of the moving rod. Suppose $v = \frac{C}{2}$; then the entire distance the light traveled from B2 to A2 will be $\frac{2R_{A1B1}}{3}$.

Can we compare the time spent by the moving rod to travel the distance $2R_{A1B1}$ to the time that the light spent traveling $\frac{2R_{A1B1}}{3}$?

We will use the four-dimensional event world line mapping method to express Einstein's moving rod systems where light is traveling in the same direction as the moving rod.

The method to study the light moving in the opposite direction to that of the rod is the same as above. We will not discuss in detail nor draw the relevant diagram.

Since we are considering only the X-axis, the Y-axis and Z-axis are ignored. Four- or five-dimensional, even higher-dimensional, event timeline mapping methods will be discussed in a later chapter.

For clarity, we first draw a figure for each step and analyze it, and then sum up these stepwise results so we can compare the stationary rod system to the moving rod system.

Each graph consists of four major parts. The top is the X-axis, to record the locations of all subjects and the distances traveled. Next is the static rod system analysis diagram. Below that is analysis of the moving rod system, and the last is the timeline of the four-dimensional object-event diagram. Each part is separated by a dashed line.

However, an important question arises: What are the steps? We need to pick some important representative moments at which to do the analysis. Meanwhile, in order to be intuitive, we need to make an image to express these moments. So, we need make the selection of which time instances there will be representative. What kind of figures do we need to draw? How many? This is what is required to

Part I A 4th grade math quiz that has been fooling human a century

constitute an effective analysis and relatively simple model to study the problem.

I want to emphasize that in the scientific research, this is the key time to affect the overall findings. After the decision is made, the rest is just number crunching. In scientific research, research direction and reasonable and effective experimental design are the most important parts of success, rather than dazzling mathematical derivations.

Here, I designed in accordance with the following methods and analysis. Before beginning the analysis step by step, I did not know what the outcome would be. I just thought that in Einstein's paper, in the negation of simultaneity, his reasoning seems insufficient and it seems he came to his conclusions too hastily. So I wanted to have a closer look, not using the velocity superposition, but by directly checking the system step by step. However, after analysis of several images, I discovered the problem. Then I starting dig out Einstein's paper in depth, and the result is got the ideas for this chapter.

First of all, because the moving rod system is more complex, I decided to focus on analyzing the moving rod system. The static rod system is just built according to that of the moving rod system.

Now, in the moving rod system, which movements are worthy of our concern?

Usually, in systems analysis, the starting state of the system is the one we must make clear. This determines the first moment that we start thinking about and designs the corresponding first system analysis diagram.

Then we look for those special detail moments within it, or the movements at the turning points. Here, it is the end of the rod. This determines the second moment that we will address and its corresponding system analysis diagram. It is the moment that the light departs from the starting rod end and travels a stationary rod length.

So, what's next? What is the change in the terminal state of the moving rod system in the second graph as the system moves? The positions of the rod ends A2 and B2 have been changed. We take the position the light reaches B2 as the third moment of attention, and draw the corresponding third subsystem analysis diagram.

So we have the three research moments of static and moving rod systems, but the third system analysis moment and diagram cannot be determined yet, and we won't be able to until we finish the analysis on the second moment and diagram.

After the third moment and diagram, should we continue our analysis for more moments and diagrams? We can't determine it at this moment. That is to be determined after we get the analysis result of the third moment.

Now let's look at the first systematic analysis on the starting moment and its figure.

System Analysis on Einstein's Rod System - Initial Diagram 1

The following Einstein initial system analysis, Figure 1.10, is obtained by using Einstein's "in thought" grouping: the four independent objects are grouped in the schematic diagram 1 of the initial state of the system of the basic model we constructed in Figure 1.8.

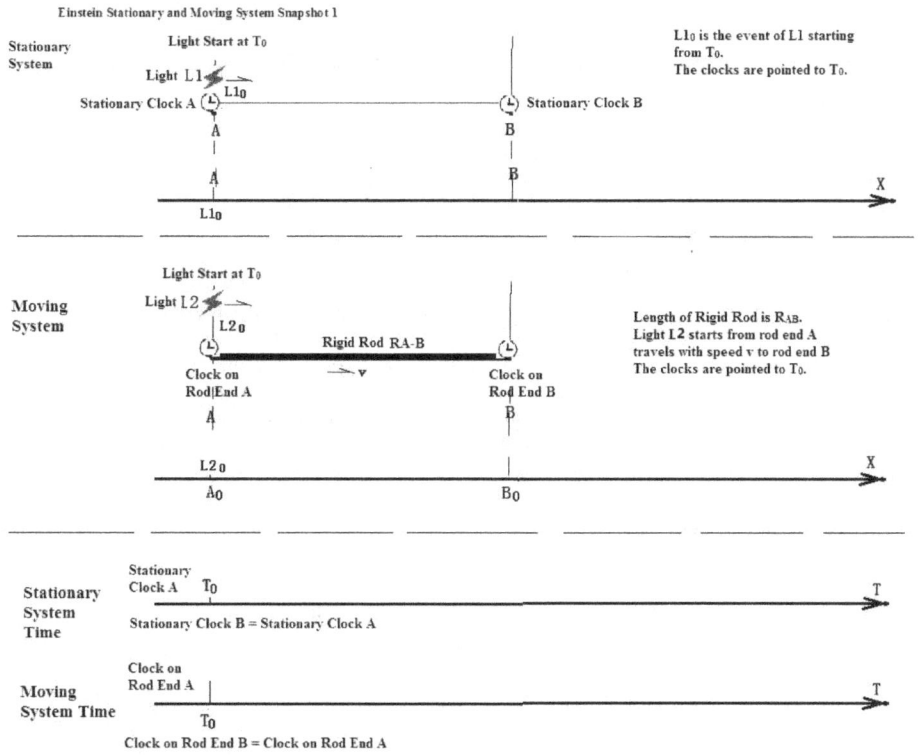

Figure 1.10 Einstein system analysis diagram 1.

The static system is composed of a static rod and a laser L1, and the event that the light L1 starts to move at the beginning time T_0 is denoted as $EL1_0$.

The moving system is composed of a movable rod and a laser L2. The event that the movement of the light L2 starts at the initial time T_0 is denoted as $EL2_0$,

Part I A 4th grade math quiz that has been fooling human a century

and the event that the moving rod R2 starts moving is referred to as $ER2_0$.

The auxiliary graph elements have a one-dimensional coordinate system X-axis located at the top of the graph, and a time T-axis located at the bottom part of the graph.

The groups are separated by dashed lines.

We call the event set E_0 ($EL1_0$, $EL2_0$, $ER2_0$) for all the events happening for all moving objects in Einstein's system analysis.

Because the speed of light L1 and L2 are the same, as long as they start at the same time and stop at the same movement, then they can be regarded as the same light in motion.

In the event of $EL1_0$, $EL2_0$, and $ER2_0$, all the clocks (static clock T_{A1}, static clock T_{B1}, moving rod end clock T_{A2}, and another moving rod end clock T_{B2}) are pointing to the same time T_0.

This is relatively simple, and we only need consider that the various factors are properly expressed.

System analysis on Einstein's rod system - diagram 2

We will choose the moment that both L1 and L2 are arriving at point X_1 to draw the system analysis Figure 1.11 shown the system state where L1 travels to the static rod end B1 and X_1 on the X-axis, as well as T_1 on the time T-axis.

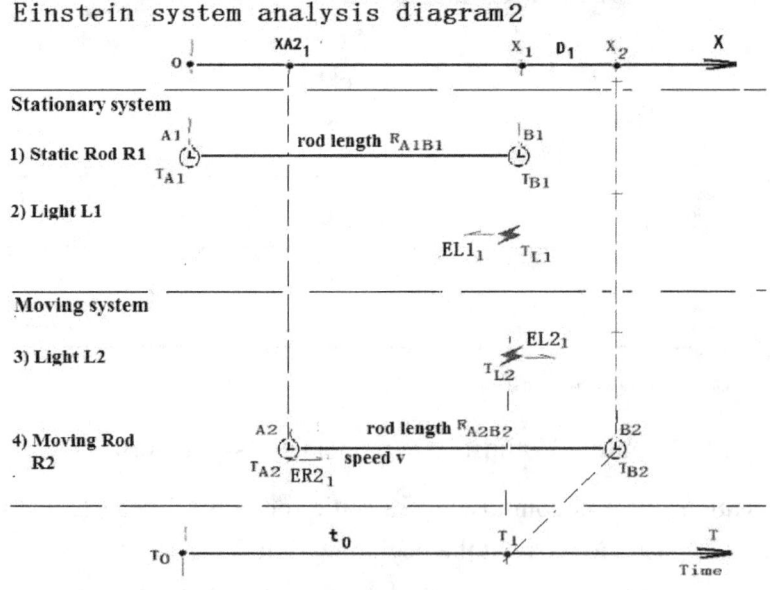

Figure 1.11 Einstein system analysis diagram 2

We refer to the moving objects of all the systems in above Einstein's system analysis Figure 2, the set of moving object events, as E_1, which contains E_1 ($EL1_1$, $EL2_1$, $ER2_1$).

The light L2 also runs to a position X_1 on the X-axis, same as the position of rod end B1, and at a time T_1 on the time T-axis.

In this case, the distance traveled by the light is L1 traveled from A1 to B1 in the static rod system, which exactly the same as the distance between A2 and B2 passing through, and is the same length as the rod length R_{A1B1}.

Thus, it can be calculated that the time of the light traveled through distance R_{A1B1} is $t_0 = \dfrac{R_{A1B1}}{C}$.

The light L1 in the static rod system travels from the starting end A1 to the other end B1, which on the X-axis is from the origin 0 to the X_1 point. Then L1 immediately reflects back to the A1 rod end, which is indicated in the figure by $EL1_1$ event.

The static rod did not move and is still in its original place.

But the position of the moving rod has been changed. The distance traveled by the light L2 in the moving rod system is also indicated on the X-axis from the origin 0 to the point X_1. It should be noted that the moving rod is moving with speed v. At the moment the moving rod end A2 is left of the origin of 0 and rod end B2 is left of X_1, it has been moving forward along the X-axis at speed v for t_0 time. The distance it advances is represented by the X-coordinate as the position from X_1 to X_2. This can be calculated in this way: the moving rod front end A2 has been moved from X_1 to the location of the X_2, so

$$X_2 - X_1 = v * t_0 = v * \dfrac{R_{A1B1}}{C}$$

In other words, L2 has not yet reached its destination – the other end of the moving rod B2. At this point, the moving rod end A2 moved to $XA2_1$ and the moving rod end B2 to X_2.

We note that starting from here it is different from the analysis of Einstein's original rod systems. Light L2 needs to continue moving forward to the moving rod end B2, and B2 end itself also continues to move. Our system analysis needs to be continued.

In the figure, we use $EL1_1$ to indicate that the light L1 arrives at static rod end B1 denoted as X_1 and immediately returns to X-axis origin 0. The $EL2_1$ indicates that the light L2 arrives at X_1 but not at moving rod end B2, and still

Part I A 4th grade math quiz that has been fooling human a century

continues the movement toward the rod end B2 along the X-axis.

$ER2_1$ is used to represent the event that the rod end A2 reaches $XA2_1$ and the rod end B2 reaches X_2, and B2 continues moving forward along the X-axis.

In the event of $EL1_1$, $EL2_1$, and $ER2_1$, all the clocks (static clock T_{A1}, static clock T_{B1}, moving rod end clock T_{A2} and another moving rod end clock T_{B2}) are pointing to the same time T_1, L1 arrived rod end B1, but L2 didn't arrive at B2.

Because L2 hasn't arrived the moving rod end B2 (B2, at present time, is located at X_2), the system analysis needs to be continued.

At this time, according to the observers that are arranged by Einstein, those observers at the rod ends found that the local clock are at different positions but all clocks are indicating the same time. This is because, with the movement of the moving rod, B2 has left from X_1 with the moving speed v from X_1 forward a distance to X_2. How to calculate this distance and the time used has been determined in the above analysis.

Now we are able to draw the following conclusion: the observers arranged by Einstein at the static rod and moving rod end find that the times they are observing are same, but the light locations are not. This is because L1 has reached the rod end B1, so the rod end clock time, indicated by t_A, has been obtained—that is, $T_1 = T_0 + t_0$; but L2 although pointing the same time as L1, has not caught up with moving rod end B2, so the observer at the moving rod end B2 has not yet begun his clock observation.

Up to here, the analysis has told us that the speed of light will not change. The clock times do not change. But Einstein's observation times (the moment the light reaches the end of the rod) for static rod system are different from the observation times of the moving rod system, because the light L1 and L2 traveled different distances at the same speed from starting time to different observation time, the observers on static rod ends and the observers on the moving rod ends are not observing the clock at the same time. At the observation moment defined by Albert Einstein, the moment T_1 at which the stationary clock is observed is also the moment the moving clock points to. But T_1 is only the observation moment of the stationary rod end clock, not the defined observing moment of the moving rod end clock, since, according to the rule of Einstein, only the observer of the stationary system observes the clock. The observer of the moving system has not begun to observe. L2 has to continue to travel a further distance until L2 catches up to the moving rod end B2, then the observer at B2 begins to observe.

So, at the moment T_1, we can already conclude that the clocks are always synchronous, but because the distance between the static rod and the moving rod at Einstein's prescribed observation time is different, the observed times of the static rod and the moving rod end clocks are different; that is, **they are in different locations having covered different distances, thus the different times observed because of the observation rule defined by Einstein.**

The next thing we want to research is: what causes the observer on the moving rod end to observe a different time while the rod is moving in the same direction as the moving rod? Is the clock itself in disarray, or, as the above analysis indicate, does the different observation times causes the different observed results?

But before starting the research mentioned above, we will try to complete the experiment that the light L2 catches up with the moving rod end B2.

To do this, we have to continue to examine the system in order to know when L2 catches up with B2, so as to know the clock time observed, and have a more comprehensive understanding of the moving rod system where the light moves in the same direction as the rod.

So we know how to draw the third system analysis figure, which is that L2 runs to the current location of B2; and B2, at the same time, also moving forward to a new location.

For the convenience of the following analysis, we first calculate the distance D_1 of the moving rod B2, moving from X_1 to X_2, from the starting event $EL2_0$ to the event $EL2_1$, and that is the distance the rod moves forward for t_0 time with speed v in the X direction. It is clear that the distance D_1 between X_1 and X_2 can be calculated as follows:

$$D_1 = X_1 - X_0$$
$$= v * t_0$$
$$= \frac{v * R_{AB}}{C}$$

System analysis on Einstein's rod system - diagram 3

We chose the moment that the light L2 arrived at the location of rod end B2 in diagram 2 above to draw the system analysis diagram 3 below.

Part I A 4th grade math quiz that has been fooling human a century

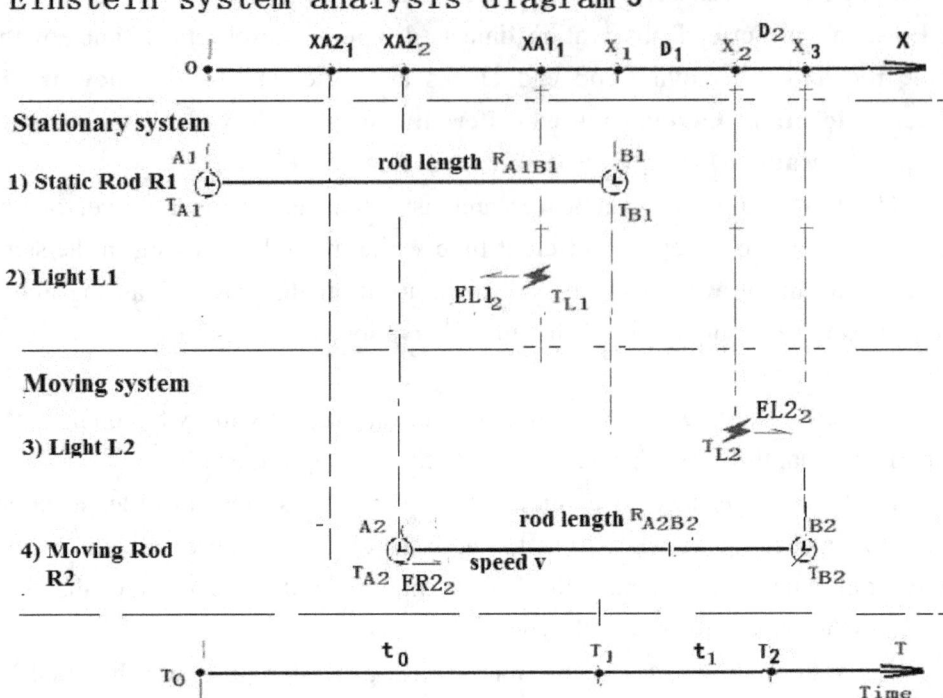

Figure 1.12 Einstein system analysis diagram 3

We name the object event set E_2 ($EL1_2$, $EL2_2$, $ER2_2$) for all of the interested moving objects of the rod systems in Einstein's system analysis diagram 3.

We used $EL2_2$ to represent the event where the light L2 departed from X_1 and arrive at X_2.

In the moving rod system, when the light L2 arrives at X_2, the movable rod end A2 is not at the position $XA2_1$, as it was in diagram 2, but has moved to the position of $XA2_2$ in diagram 3 with speed v. The moving rod end B2 now arrives at the point X_3 from X_2, and the distance traveled by is denoted by D_2.

Then, how big is the distance D_2 from X_2 to X_3? In fact, this is calculating the distance that the light travels within time t_1 and velocity v along the X-axis direction, where t_1 is the time used by the light L1 to travel through D_1 in system analysis diagram 2. Obviously, the distance D_2 from X_2 to X_3 can be calculated as:

$$t_1 = \frac{D_1}{C}$$

$$D_2 = X_3 - X_2$$

$$= v * t_1$$

$$= v * \frac{D_1}{C}$$

$$= v * \frac{v * R_{A1B1}}{C}$$

$$= R_{A1B1} \left(\frac{v}{C}\right)^2$$

At the same time, as systems analysis shows in Figure 1.12 above, in the stationary rod system, the light L1 reflects from B1 toward A1 and travels from X_1 to $XA1_1$ for distance $X_1 - XA1_1 = D_1 = R_{A1B1}$. The event of L1 traveling for D_1 is named $EL1_2$.

Rod end B2 has moved away from X_2, traveled distance D_2, and reached X_3.

At this moment, the stationary rod system and moving rod system clocks all are pointing to $T_2 = T_1 + t_1$. That is, $T_{A1} = T_{B1} = T_{A2} = T_{B2} = T_{L1} = T_{L2} = T_2$. There is no time difference between these clocks.

Thus, L2 also needs to start from X_3 to reach its final destination – the rod end B2 that is moving with speed v.

After analysis of the system diagram 3, we found that the patterns of system diagrams 2 and 3 were almost repeated. Putting the above analysis together, and searching our stored knowledge in our memory, we discover that light moving from one end to catch up to the other end destination for the moving rod system has the same movement pattern as the race between the tortoise and the hare that was introduced at the end of this book in the presentation example of catch-up mode in the section of prior knowledge. They are exactly the same. So we know how to solve the moving rod problem.

Here the knowledge using is more than middle school mathematics, and is more complicated than the mathematics Einstein used in the relevant section of our discussion. The expression is relatively simple, but a little complicated in calculation. We need to analyze and calculate in accordance with the methods used to solve an infinite sequence. You can go online to check the calculation

Part I A 4th grade math quiz that has been fooling human a century

method of the sum of an infinite sequence to get a detailed answer. This is the only few really difficult mathematical formula in this book that requires more than high school knowledge of math. If it's too thorny for you, feel free to skip it.

First, from previous analysis, we know that in the moving rod system, the light L2, after any distance D_i and time t_i spent, can be calculated as follows:

$$D_i = R_{A1B1} * \left(\frac{v}{C}\right)^i$$

The whole distance that the light L2 travels from rod end A2 to the destination end B2 in the moving rod system can be calculated as follow:

$$D = D_0 + D_1 + D_2 + \ldots + D_i + \ldots$$

$$= R_{A1B1} + R_{A1B1} * \left(\frac{v}{C}\right) + R_{A1B1} * \left(\frac{v}{C}\right)^2 + \ldots + R_{A1B1} * \left(\frac{v}{C}\right)^i + \ldots$$

$$= R_{A1B1} * \left(1 + \frac{v}{C} + \left(\frac{v}{C}\right)^2 + \ldots + \left(\frac{v}{C}\right)^i + \ldots\right)$$

$$= R_{A1B1} \sum_{i=0}^{\infty} \left(\frac{v}{C}\right)^i \qquad (4)$$

From equation (4) above, we know that the value of D is mainly determined by the velocity v of the moving rod.

Now it's time to apply our research and calculation results. Based on the above results, let's make a comparison of our research results against Einstein's model and his conclusion, and find out what Einstein missed in his research paper. What kind of mistake did he make?

Einstein's RELATIVISTIC MODEL CANNOT BE ESTABLISHED BECAUSE although the mathematical calculations in his original publication are correct, but the physical interpretation of the equations are incorrect

We give moving rod velocity v an easy calculation value, and substitute it into the formula (2) and (4). This gives us an intuitive way of specific example to understand the problem.

The selected value is $\frac{C}{2}$, half the speed of light. We're choosing it here just because it is computationally convenient, no other reason. Because the speed of light is the absolute speed of the universe (up for debate, but let's use this widely accepted concept here), you can choose any value less than C for v, of course; it will be more trouble to calculate and to explain, but the results will be the same.

If the velocity v of rigid rod is half the speed of light, that is,

$$v = \frac{C}{2},$$

then substituting it into formula (4) will result in

$$D = 2 R_{A1B1}$$

If the velocity v of a moving rigid rod is half the speed of light, light L2 ultimately travels twice the distance of L1 in the stationary rod system. That is, the distance light L2 travels is $2R_{A1B1}$ in the moving rod system. L2 needs pass double rods length to finally catch up the moving rod end.

At any time during the catch-up process, the clocks in the stationary rod system and the moving rod system are pointing to the same time. While $v = \frac{C}{2}$, the light L2 in the moving rod system travels 2 R_{A1B1} distance in $\frac{2R_{AB}}{C}$ time.

And because, in the stationary rod system, the Light L1 is synchronized with L2 during this time period, L1 has traveled a distance $2R_{AB}$. L1 not only ran from point A to point B, but also returned from B to A. Using Einstein's notation, the time the stationary rod system actually used is:

$(t_B - t_A) + (t'_A - t_B)$

Instead

$(t_B - t_A)$

Note that:

In fact, the problem can be revealed from the following calculation, and also can obtain the same conclusion from Einstein's formula (2). This shows on the one hand that the scientific principles are interlinked, and mathematics is always accurate; on the other hand, it also shows that our understanding of Einstein's model is correct; it is mainly from the perspective of physics to analyze and solve problems.

We substitute $v = \frac{C}{2}$ into (2), Einstein's base model of setting up the relativity concept, to obtain:

Part I A 4th grade math quiz that has been fooling human a century

$$t_B - t_A = \frac{R_{AB}}{\frac{C}{2}} \tag{5}$$

But according to the physical principle and formula (4), this time period for the moving rod system should be calculated as follows

$$t_B - t_A = \frac{R_{AB}}{\frac{C}{2}} = \frac{2R_{AB}}{C} \tag{6}$$

Einstein used formula (5) to derive the relativity concept, the physical meaning says that light travels moving rod length R_{AB} with velocity of half-light speed $\frac{C}{2}$.

But actually, according to our physical analysis of the moving rod system, this time period should be calculated with formula (6), the light traveling $2R_{AB}$ with normal light speed C.

Obviously, in (5) and (6), **the mathematics are the same** and also produce the same results.

With regards to the application of (5), although the math result is correct, **the physical sense is wrong**. Light cannot travel with half of light speed, and the distance traveled by light in the moving rod system is not R_{AB} but $2R_{AB}$. That is correctly shown in (6). Einstein directly applied the mathematical formula of speed superposition on the moving rod system, which masked the details of the movements of the light and rod. Thus, he compared the time at the stationary rod end in the stationary rod system to the time of light arriving at the rod end of the moving rod system, which are totally different in reality but were considered to be the same in Einstein's calculation. Accordingly, he denied the absolute nature of simultaneity, and based on this, drew out the wrong conclusion of the relativity concept.

Once again: while the same math results were obtained from the computing, the physical meaning of formula (5) that was used by Einstein was completely wrong. The speed of light for the moving rod can only be C instead of $C - v = \frac{C}{2}$. The fact that the speed of light is always the same is proposed by Einstein himself, and it is now a generally accepted view of the physical world.

While making the application of formula (6), the true distance that the light traveled clearly manifested itself, so we will not compare the movements of stationary rigid rod ends to that of the starting and end points of the moving rod system. Obviously, (t_B - t_A) in the moving rod system is $\frac{2R_{AB}}{C}$, but in the stationary rod system it is (t_B - t_A) is $\frac{R_{AB}}{C}$. After understanding the different physical meaning of t_B in stationary and moving rod systems, we will not conclude that we should deny the absolute nature of simultaneity, which finally led to the emergence of a relativistic time-warped mistake.

However, starting from formula (2), people will use formula (5) and will certainly incorporate the longer distance light travels into the speed change, arrive at the same mistake that Einstein made using the wrong physical meaning, and draw out the wrong conclusions, and therefore lead to the time-warped "moving clocks are slower" paradox and "moving meter stick is shorter" disaster.

Note that all the above data is calculated from a rod velocity set at half the speed of light in the moving rod system. If rod speed is not just half the speed of light, then the above data calculated is different, but the reason is exactly the same.

The "slower" time of "moving clocks are slower" and the "shorter" length of "a moving meter stick is shorter" in Einstein's relativity theory are actually hidden in the "moving" velocity v. This is the conclusion from the comparison of different physical meaning between formula (5) and (6).

Velocity v is an intermediate concept, and is determined by the distance and the time taken across this distance. In mathematics or arithmetic problems, we have many direct and simple problem-solving approaches. However, these calculations omit the intermediate step in understanding the physical processes and will bring confusion when used in solving physics problems.

In the calculation of the movement of light from one end to the other end of the moving rod system, Einstein used the synthesis of movement and speed, which certainly in the calculation is correct, but such a direct and concise contrast with the stationary rod system – "simple" math but wrong physical meaning – leads to the incorrect concept and the conceptual error conclusion. Einstein was a master of mathematics and physics, and in dealing with such an extremely simple problem of light travel in moving rod motion, without thinking, he used the most straightforward solution, but this simple approach cannot be directly used to compare a stationary rod system to a moving rod system and then, from it, extended to a new physical meaning.

Part I A 4th grade math quiz that has been fooling human a century

Analysis of formulas (2) and (2.1) again:

$$t_B - t_A = \frac{R_{AB}}{C-v} \quad (2)$$

$$t'_A - t_B = \frac{R_{AB}}{C+v} \quad (2.1)$$

Formula (2) tells us that the light starts from moving rod end A at the speed of (C - v). At the time that the light is catching up with the moving rod end B, the rod has traversed distance R_{AB}. Formula (2.1) tells us that the light is in the opposite direction at the speed of (C + v). To catch up with the moving rod end B, the light also passes R_{AB} distance. Now the light travels the same distance at different speeds, but does not satisfy $t'_A - t_B = t_B - t_A$ as in (1) of the stationary rod system. So the different observers at different rod ends see different clock times. But according to Einstein's relativity definition, these clocks should indicate the same time. Einstein himself stipulated that the speed of light is unchangeable. The speed of light cannot be changed because of the relative movement of a rod. Perhaps the clock observers on the different rods will "feel" that the speed of light has changed, but in fact, this is only the illusion of the observer. It is also impossible to change the time indicated by the clock attached to the rod. Think of the rod as a boat. If the passengers see lightning in the sky, will the time indicated by the clock on board change? (The reader can also discuss or even write a paper on the situation in which a wheel rotating at the speed of light rotates forward on the rod and the friction between the wheel and the rod is perfect. But we do not discuss this situation in this book, in which the light moves independently and has nothing to do with the movement of the rod.)

Since the speed of light is constant and is the highest speed, the two formulas can be changed in such a way:

Let $v = kC$, $(0 < k \leq 1)$. Then

$$t_B - t_A = \left(\frac{R_{AB}}{\frac{1-k}{C}}\right) \quad (7)$$

$$t'_A - t_B = \left(\frac{R_{AB}}{\frac{1+k}{C}}\right) \quad (7.1)$$

in which

(7) = (2); that is, (2) and (7) are the same.

(7.1) = (2.1); that is, (2.1) and (7.1) are the same

The physical meaning of equations (7) and (7.1) says the light with the unchangeable speed C runs after the moving rod with speed v, traveling $\frac{R_{AB}}{1-k}$ distance to catch up to the other end. In reverse direction, movement passes $\frac{R_{AB}}{1+k}$ distance to catch up with the other end.

The same speeds but different distances traveled, of course, are not occurring over the same time. It is obviously extremely unreasonable to ask them to be the same. (2) and (2.1) are typical examples of correct mathematics but wrong physical meaning.

The observer on the moving rod sees an illusion, because (7) != (7.1) is observed. Einstein argued that (2) = (2.1) in the moving rod system, the same as the formula (1) in the stationary rod system. But this is entirely Einstein's own illusion, created by himself after he gave and used a variety of definitions and mixed unclear symbols, creating chaos.

Analysis of the Light Moving in opposite direction to the movement direction of the Rod

We did a detailed analysis of when the light L2 moves in the same direction as the moving rod R2, so we can know the specific method and model of the system analysis, get the satisfactory results, and draw the conclusion: Einstein defined the relativity principle and gave the observation rule so that when the light reaches the rod end, the observer there starts observation. This causes the observers of different rods to observe at different times after they traveled different distances, and thus got the different observing results. But we cannot use such results to deny simultaneity and establish relativity, because, in this process, all the clocks are still indicating synchronous time.

So, can we apply the same analysis to the case in which the light moves in the same direction as, or in opposite direction to, the movement direction of the rod? Shall we get the same conclusions about the distance traveled and the time used in such a contrasting observation?

We use Figure 1.13 to observe the contrast of the system state for the light and moving rod move in the same direction and for the light and the rod moving in opposite directions.

Part I A 4th grade math quiz that has been fooling human a century

In the upper box of Figure 1.13, a light is between two moving rods traveling in the X-axis's positive direction; the upper moving rod moves in the same direction with speed v. In the lower box, the rod also moves with velocity v but in the opposite direction of the light. They start at the same time. In this way, we modified Einstein's model (light moves to the rod end and then back to the starting end,) but did not modify the calculation essence of the same direction and reverse movement, which gives us an intuitive contrast and makes it easier to see where the problem lies.

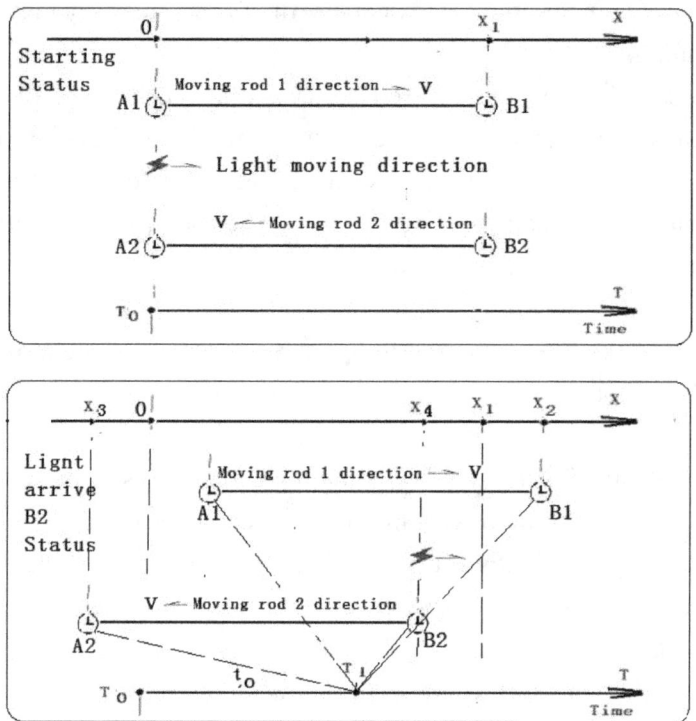

Figure 1.13 The light is traveling relative to two moving rods, with one rod in the upper frame moving in the same direction as the light, and another rod in the lower frame moving in opposite direction against the light. In the upper frame, the light and the rods start moving at the same time; in the lower frame, the light traveled some distance and is reaching the rod end B2.

The lower box of Figure 1.13 shows the light arriving at the rod end B2 moving in the reverse direction. At this point, in the upper box, the moving rod end B1, which is traveling in the same direction as the light, is still in front of the light. Since the light reaches rod end B2, which moves in opposite direction to the light, the observer at B2 records the time indicated at this moment; but the observer at the rod end B1, which moves in same direction as the light, does not see the arrival of the light and thus did not record the time at this moment, the same time value that all clocks are indicating. The clocks at all the rod ends are indicating the same time, but the observer, in accordance with Einstein's observation requirements, does not make an observation of all of them at this time. Since the observation time is the moment when the light reaches the rod end, the observer at the rod end B1 has not yet seen the light catching B1.

When we focus on the positions of B1, B2, and the light, we know that both light and B2 are at X_4, but B1 is at X_2. The four rod ends are indicated at the time T_1. But at this point, only the observer at B2 was recording the clock time because the light had reached the position of the rod end where the observer is located. The other three observers did not observe and make a record at this time.

The rest of the drawing, analysis, and calculation process is the same as before. I will not waste book space here, leaving it to the interested readers themselves to complete.

We can see from these analyses that (2) = (2.1) is purely due to the illusion that Einstein's observed times are actually different. The observers were observing clocks at different times, following Einstein's definition that clock observing will only happen when the light is arriving at the rod end. With movement in the same direction or reverse, the light traveled different distances to the rod ends, and the times for the light arriving at the rod end are different.

The speed of light has not changed; the distances traveled are different, so the amounts of time, of course, are different. This, on the contrary, from another angle, proves the absoluteness of simultaneity: traveling different distances takes a different amount of time; traveling the same distance takes the same amount time.

This also proves that Einstein's relational reasoning process of denial of simultaneity and, thus, the construction of relativity is not correct. It is the use of velocity to confuse the distance and time. As long as we adhere to Einstein's own principle that the speed of light can't be changed, then (5) is a formula that have the correct mathematical calculation but the wrong physical meaning. Using the results of the calculation to make relative comparison is unreasonable; the conclusions drawn will be wrong!

The basic concepts are not correct, then can the complex theory derived from it be correct? So the increasingly complex mathematical derivations based on the establishment of the concept of relativity in the later parts of "On Electrodynamics of Moving Bodies" are meaningless and can be ignored.

Part I   A 4th grade math quiz that has been fooling human a century

5. WHY IS THE RELATIVITY PRINCIPLE DEFINED IN EINSTEIN'S PAPER UNREASONABLE?

A reasonable definition in simple and clear language is needed to explain the nature and character of an object or event.

Einstein follows four steps to deny simultaneity and establish relativity. They are: 1. He defines the synchronization in the stationary system; 2. He gives the relative principle definition; 3. He applies the definition to the rod systems; 4. He proves that the relativity concept is established, i.e., the concept of simultaneity is not established. Following are the details:

-- (The following italicized texts are copied from "On the Electrodynamics of Moving Bodies")--

I.   *Define the synchronize of in stationary system:*

In accordance with definition the two clocks synchronize if

$t_B - t_A = t'_A - t_B$ （1）   (the number of the formula is added by author)

II. Define the principle of relativity *:*

"§2 On the Relativity of Lengths and Times

... based on the principle of relativity and on the principle of the constancy of the velocity of light. These two principles we define as follows The laws by which the states of physical systems undergo change are not affected, whether these changes of state be referred to the one or the other of two systems of co-ordinates in uniform translatory motion.

III.   Applying the above definition in the moving rod system

We imagine further that with each clock there is a moving observer, and that these observers apply to both clocks the criterion established in §1 for the synchronization of two clocks. Let a ray of light depart from A at the time t_A, let it be reflected at B at the time t_B, and reach A again at the time t'_A. Taking into consideration the principle of the constancy of the velocity of light we find that

$t_B - t_A = R_{AB} / (c - v)$ （2）   (the number of the formula is added by author)

$t'_A - t_B = R_{AB} / (c + v)$ (2.1)

IV. From the observe results of the observer

Where R_{AB} denotes the length of the moving rod—measured in the stationary system. **Observers moving with the moving rod would thus find that the two clocks were not synchronous, while observers in the stationary system would declare the clocks to be synchronous.**

V. The proof proves that the concept of relativity is established, that is, the concept of simultaneity is not established

So we see that we cannot attach any absolute signification to the concept of simultaneity, but that two events which, viewed from a system of co-ordinates, are simultaneous, can no longer be looked upon as simultaneous events when envisaged from a system which is in motion relatively to that system.

-- (The above italicized sections are copied from "On the Electrodynamics of Moving Bodies") –

Above excerpts the core from Einstein's paper of §1 and §2, we can clearly see the logic problem in it:

The synchronization in static system was defined in I; then based on the relativity principle defined in II, apply formula (1) in III, to require (2) = (2.1). Since they are not equal in IV, we get the conclusion in V.

The question is whether the relative definition in II is reasonable. In the absence of proof of the rationality of the relative principle definition before its application, extended the synchronize equation (1) obtained from static rod system to the moving rod system, requires that the synchronization of the static rod system can also be applied to the moving rod system. Now, obviously, (1) cannot be extended to the moving rod system as common sense; that is, (2) != (2.1) is the basic knowledge, how can he from here to get the conclusion that absoluteness of simultaneity is wrong (relativity is correct), is this not too hasty?

Under the definition of integrity requirements, the object and the epitaxy should be equal for all aspects of the concept and for full disclosure of the full meaning. For example, a triangle with two equal sides is an isosceles triangle. On the other hand, isosceles triangles are triangles with two equal sides.

Now we also simply change the order of Einstein's definition application: first, derive (2) and (2.1) from the moving rod system III, and then try to extend them to I. You cannot find that it can be applied there. The reason this cannot be extended is that the definition of II does not hold. Einstein's own proof proves that the "relative principle" in II is irrationally untenable.

In fact, before apply the relativity principle, we already knew (2) ! = (2.1). It should not possible to try to generalize this inequality to equation (1) by the relativity principle.

It is reasonable to require that the Einstein's relativity principle definition should be a general principle that be possible to apply in any system before

Part I   A 4th grade math quiz that has been fooling human a century

applying it. But in fact in any systems having different moving speeds are not correct to use it. By using his own defined relativity principle, make the systems having different movement to become the same, obviously is not a scientific way.

In fact, correct logic of §2 should be such that: based on the relativity principle, got the wrong results after applied the relativity principle, then proved the relativity principle is wrong. Following correct scientific way, the conclusion in IV should therefore be that "we cannot use the relativity principle because its definition is not reasonable" rather than that "we cannot attach any absolute signification to the concept of simultaneity".

Let us compare the following "The Principle of Similar Law of Fruits" to intuitively understand the point of view discussing here.

The Principle of Similar Law of Fruits

1. Definition: Two oranges, if they look like the same, tastes the same, one orange taste is A, another tastes A1, then there will always be:

$$A = A1 \qquad (1)$$

2. Definition of Principle of Same Taste Fruits: the knowledge gained from the two fruits that look the same can be used to other two fruits that look the same.

3. There is a yellow apple and a yellow pear. According to the Principle of Same Taste of Fruits, extending the (1) to the pear and apple,

$$B = \text{taste of the yellow pear} \qquad (2)$$
$$B1 = \text{taste of the yellow apple} \qquad (2.1)$$

4. However, according to the conclusion of the tasters, the taste of the yellow pear and the taste of the yellow apple are not the same.

5. Therefore, it is concluded that the two apparently looks similar fruits do not taste the same.

Where is the problem of The Principle of Similar Taste of Fruits?

The "2. The definition of the principle of similar taste of fruit" is wrong. The tastes of yellow pear and yellow apple are not the same. Applying the principle of same taste of fruits to two pears or two apples is obviously wrong. The absurd result suggests that the application of the principle of same taste of fruits is wrong, rather than "5. It follows that the two apparently identical fruits do not taste the same."

The movement of light between the stationary rod ends is the inherently different from the light moving between the ends of the moving rod. This basic truth no need to be proved or apply relativity principle. Einstein defined relativity principle and the principle of the constancy of the velocity of light by himself, and used the self-defined principles to extend the static light movement time of (1) to the moving rod light time (2) and (2.1).

From (2) and (2.1), it should get the conclusion that the definition of relativity principle is wrong and cannot be used to extend (1) to (2) and (2.1); but absurdly, Einstein turned the conclusion to that it is proved that the simultaneity of absoluteness is not established. This miracle turn like done by God, has been turning the human thinking hundred years ah!

No wonder around to get around that in the world few people can understand the relativity concept ah.

6. IN FULL ACCORDANCE WITH EINSTEIN'S METHOD OF PROOF, USING THE RELATIVITY PRINCIPLE DEFINED IN HIS PAPER TO PROVE THAT THE CONCEPT OF RELATIVITY IS NOT CORRECT

In the previous section, we see that Einstein first gave a definition of "relativity principle" at the beginning of §2, applied this definition in subsequent proofs in this section, and then, at the end of §2, concluded that the concept of absoluteness of simultaneity is not established; that is, the concept of relativity is correct. In this approach, the definition is given first, and then the definition is applied. Finally, this definition is proved. That is obviously a wrong-headed method of circular argumentation.

The following fully modeled according to above section 5, only swapped the order of (1), (2) and (2.1). That is, in accordance with the principle of relativity extending from (2) and (2.1) of moving rod system to the (1) of static rod system, we can follow Einstein's Method to prove the correctness of the absoluteness of simultaneity, i.e. to deny relativity concept:

I. Define the not synchronize of in moving rod system:

In accordance with definition the two clocks not synchronize if

$$t_B - t_A = R_{AB} / (c - v) \qquad (2)$$
$$t'_A - t_B = R_{AB} / (c + v) \qquad (2.1)$$

II. Define the principle of relativity :

"§2 On the Relativity of Lengths and Times

Part I   A 4th grade math quiz that has been fooling human a century

... based on the principle of relativity and on the principle of the constancy of the velocity of light. These two principles we define as follows. The laws by which the states of physical systems undergo change are not affected, whether these changes of state be referred to the one or the other of two systems of co-ordinates in uniform translatory motion.

III.   Applying the above definition in the static rod system

We imagine further that with each clock there is an observer, and that these observers apply to both clocks the criterion established in §1 for the synchronization of two clocks. Let a ray of light depart from A at the time t_A, let it be reflected at B at the time t_B, and reach A again at the time t'_A. Taking into consideration the principle of the constancy of the velocity of light we find that

$$t_B - t_A = t'_A - t_B \qquad (1)$$

IV. From the observe results of the observer

Where R_{AB} denotes the length of the moving rod—measured in the stationary system. **Observers moving with the moving rod would thus find that the two clocks were not synchronous, while observers in the stationary system would declare the clocks to be synchronous.**

V. This proves that the concept of relativity is not established, that is, the concept of simultaneity is established

So we see that we can attach any *absolute* signification to the concept of simultaneity, but that two events which, viewed from a system of co-ordinates, are simultaneous, can be looked upon as simultaneous events when envisaged from a system which is in static relatively to that system.

The above proving process and the original I to V steps basically did not change, except the exchange of formula position, and "is" to "not", or removed "not". So we got the opposite result from Einstein's paper.

Interesting? In strict accordance with Einstein's proving method, the principle of relativity can be used casually, can be used to prove relativity concept is right, also may be used to prove relativity concept is wrong.

7. EINSTEIN'S PAPER SHOULD NOT HAVE EXTEND THE USE OF THE LOCAL TIME AS THE WHOLE SYSTEMS TIME

In the process of Einstein's proof of the relativity concept, the time calculation for the static rod end was randomly extended to the moving rod end using the same mathematical symbols: t_a, t_b to name the the time of the static rod ends and moving rod ends. This completely confused the different nature of the physical phenomenon.

For example, in the fourth-grade "sailboat in running water" math problem, the case of a boat moving against the direction of running water is different from the boat sailing in still water. Over the same time period, the distances the boats sail will be different; or if they sail the same distance, the time used will not be the same for both.

But Einstein said: "Observers moving with the moving rod would thus find that the two clocks were not synchronous, while observers in the stationary system would declare the clocks to be synchronous."

This is the specific time the individual observer observes the clock at his distance while the light reaches the observer. It is not possible to equate the observer's individual observation time with the system-wide time, since other observers did not look at their clocks at their locations at that time.

In the moving rod and light system, when the observer sees light reaching his position at the moving rod end, he merely "feels" the light reaches the local rod end. But he could not extend this feeling to the static rod end by the unreasonable definition of the relativity principle given by Einstein, since the clock at the end of the moving rod passed a longer distance and the time indicated by the clock was different. **This precisely tells that absoluteness simultaneity is established, because traveling different distances will consume different times, through traveling the same distance will use the same amount of time.**

Static and moving rod systems are Einstein's "ideological" combination, and different observers are in the "ideological" set at local observation spots. But in real life, we can carry out other, more reasonable "ideological" combinations.

For example, the static rod, moving rod, and light are placed in an observation system, as in the Einstein system analysis above. We will immediately know that when the light reaches the static rod end, the light needs to move forward to catch up to the moving rod end. No matter where the clocks hang, the time indicated by all clocks will not be confused.

For another example, as we did before, when each object is a separate system and each system has its own clock, the clocks will never be confused.

Therefore, in the correct system analysis, free to merge or split objects for research, the times indicated by the clocks will not appear disorder in the system.

Part I A 4th grade math quiz that has been fooling human a century

On the contrary, in Einstein's static and moving rod systems, if we take a rod moving at different speed, add the new moving rod to the original moving rod system, and group the two moving rods traveling at different speeds, the static rod, and the light together to form a new system, then, according to Einstein's theory of relativity, the time indicated by the clock on the moving rod relative to that on the static rod, and also relative to the newly added moving rod, will be different. So, how to decide what time this clock is indicating?

8. IN THE SIMPLE MATHEMATICS, THERE IS PROFOUND TRUTH- THE PHILOSOPHY IN THE MATHEMATICAL LIMIT THEORY SAYS "NO" TO THE PRINCIPLE OF RELATIVITY

Rush of thought, its sparks sometimes inadvertently ignited by positive and negative collision, in the vast universe illuminate a breakthrough path.

The positive and inverse reasoning and proves in section 5 and 6 above give us more complex imagination. Which also contains some interesting philosophical speculation.

If we write (2) and (2.1) as the limit form as follows:

$$\lim_{v \to 0} (R_{AB} / (C - v))$$

$$\lim_{v \to 0} (R_{AB} / (C + v))$$

Putting them and (1) together, we found an interesting phenomenon:

$$\lim_{v \to 0} (R_{AB} / (C - v)) = \lim_{v \to 0} (R_{AB} / (C + v)) = t_B - t_A = t'_A - t_B$$

The above equation tells us that in the case where the moving speed of the moving system approaches zero, the time required for the forward and reverse running of the moving system is the same as that of the static system.

Thus, in the case of reaching the limit, the movement of the static and dynamic rods unified.

Does this explain the relativity of Einstein's principle has a little reasonable? No at all!

Mathematical limit thinking can be associated with dynamic and static transformation of the philosophical thinking of the various phenomena. Mathematical limit is a series of infinite data that changes dynamically according to certain laws. Static state is at the end of the evolution. The system evolution is from infinite motion to an ultimate static state; from quantity accumulation to quality change; from process to result; from infinite approximation to precision; from the diversity to the unity ...; from the relative truth to the absolute truth.

Read such a long list of the corresponding phenomenon of the mathematical limit, what is the most important do we find?

For our discussion of the problem: any dynamic mathematical limit system is certain numbers in an infinite series of changes; the end of the change is the resulted limit of the serial numbers, which are related but qualitatively different

Velocity v takes any value that is not equal to its limit value, but all the values of v changes from big to zero, that make up their final limit value.

In the appendix, "some simple prior knowledge", in the subject that the rabbit to catch up with the tortoise, before the rabbit not catch up the turtle, rabbit and turtle are in motion, after the rabbit catches up with the turtle, the movement was discontinued, as the results of the study has obtained.

The speed to take any value is not a static state, only when the speed reaches its limit, that is, when the speeds of the moving rod is really zero, the speed reaches the limit value , The movement of the rod is over.

Since the moving rod and the static rod are essentially different, then Einstein's definition of the principle of relativity, extended the law of the static system to the dynamic system, or vice versa, are not correct.

From a philosophical point of view, the definition of relativity principle, with its correctness has not proved, is a wrong definition.

This problem can also be described from another angle.

When v tends to infinity,

$$\lim_{v \to \infty} R_{AB}/(C-v)) = 0$$

$$\lim_{v \to \infty} R_{AB}/(C+v)) = 0$$

Part I A 4th grade math quiz that has been fooling human a century

Their limit values are present, but this limit is physically wrong, obviously. Because no matter how long the rod is in the equation, the limit of time for moving along the rod is zero when v tends to infinity. But obviously the time required to walk a huge space cannot be zero.

Another example is the limit of the star light propagation, which we discussed in the appendix. When the distance is infinite, the limit of the light intensity of a star is zero. But in fact we cannot accurately say that at what distance there is no stellar photon propagation. We can only know from the qualitative point of view: the distance of the stars is limited. But specific to a particular celestial body, will have to calculate in order to know how big this distance is in the end, to meet the human "invisible" requirements. If we know that the ultimate limit of sunlight is zero, so that the Sun can only illuminate the limited distance of one hundred thousand light years, it is completely wrong. By the same reason, the increase of the speed of the moving rod will eventually make the time passing through the two ends of the moving rod to zero, but cannot use a specific speed, or (2) and (2.1) extend to the limit state (1) according to the relativity principle. However, it is still not possible to generalize the extension of the static state of the limit state to the state of dynamic changing system according to the relativity principle.

This from the perspective of mathematical philosophy illustrates the limit process and the limit itself is not the same, cannot equate the process and the result. While Einstein's definition and application of the relativity principle confused the limit process and limit result, it is clearly wrong and not desirable.

Relativity and the relativity Principle

In above discussion, we focused on the ends of the rigid rod. In fact, working on any point of the rod will lead to the same conclusion. We can see that in whole process of the analysis and comparison of Einstein's stationary rod system and moving rod system, there is no time violation of the absolute simultaneity. In other words, Einstein in § 2 **would thus** conclusion in his model is wrong. According to the results obtained by the models, he cannot deny the existence of simultaneity between stationary rod system clocks and moving rod system clocks.

Also, in our real life, we rarely see practical examples of violations of simultaneity. According to Einstein's conclusion, any movement of the object, no matter what the speed of movement, will have to change the time or size (these changes led to the impossible "moving clocks are slower" paradox and "a moving meter stick is shorter" disaster). But when we take a train for thousands of kilometers or fly around the Earth, nobody's clock or watch shows any non-simultaneity. The use of GMT illustrates that the principle of simultaneity in existence on Earth is reasonable and absolute. Clocks around the world are set using the same absolute clock system and run for many years with no problem. As for time adjustments, the difference depending on the region is not the simultaneity problem Einstein described, but is due to regional adjustments in human-maintained time computing systems.

If the relative movement would cause the time change, then, at the same time moving in all directions, dozens, hundreds or even thousands of transport vehicles, like in the World War II thousands of airplanes fighting together, will cause what kind of time and space disorder? The clock on each aircraft indicated different time that would be what kind of chaos?

Relativity may exist in a certain sense, but it is not Einsteinian obfuscation of time and space, the principle of relativity. Moreover, from Einstein's models of static and dynamic rod systems, the absoluteness of simultaneity cannot be denied, nor the rationality of the relativity of temporal and spatial chaos can be proved.

In general, if two systems moving at different speeds, from one to observe the other, there is always some degree of relative motion that needs to be considered, such as calculating the velocity between them, the relative magnitude of the light received from distant planets in different time. However, within the limits of what we know and be able to consider, this relativity does not affect the simultaneity of absoluteness. On the contrary, it is with the absolute nature of the simultaneity, we know the star light from millions of light years away is issued a million years ago. This is the normal conclusion in the "relative" after traveled the different positions of the space. Also with the absoluteness nature of the simultaneity, it is possible to have the basis to talk about the relative motion between different objects.

The normal "relative" nature is based on the simultaneity of absoluteness, not on the basis of the relative observation of two different moving objects. Otherwise there will be the contradict errors and uncertainty of multiple relative systems.

Einstein's theory of relativity, which is defined in "On Electrodynamics of Moving Bodies," argues that the clocks in systems moving at different speeds, when they are "relative" to each other, pointing to different times in the system,

Part I A 4th grade math quiz that has been fooling human a century

which led to the described ridiculous inconsistent results.

Einstein's "relativity principle" is a result of thinking and only relative in thinking, is a result of imagination, rather than the actual clock pointer or spacecraft can change the length of the physical phenomenon; is the product of imagination, should not be applied in practice.

Do not confuse the relative of everyday life with Einstein's relativity principle that could change the pointing time of clock's pointer.

Answer to the questions and issues of fly rabbit, turtle, dog and watches

Before concluding this chapter, let's look back at the little fable about the dog, the flying rabbit, the turtle, and the watches at the beginning of this chapter. Now we can follow the model established by Einstein's relativity theory and get some answers.

The answer to the first question: The watch the flying rabbit is wearing is always right.

The answer to the second question: The watch the dog is wearing is always right;

The answer to the third question: When the flying rabbit catches the turtle, their watches will begin to fall out of sync.

Why?

Because according to the theory of relativity, following the model based on Einstein's theory of relativity established in §1 and §2 of his original research paper "On the Electrodynamics of Moving Bodies", the answer is thus.

Details are as follows: (see Einstein's original paper "On the electrodynamics of moving bodies" attached to the back of this book).

The first question: replace the dog with the light of stationary rod system, and the runway with a rigid rod. Because no relative reference system here either, the timing of the watch is not affected, so the watch will continue to give correct time as long as it is not damaged.

The second question: if we replace the flying rabbit with the light of moving rod system, the runway with a rigid rod, and turtle with the B-end of the moving rod, then it becomes identical to Einstein's moving rod system as in his paper. But because there is no relative reference system, the timing of the watch is not affected, so the watch will continue to give correct time as long as it is not damaged.

Third question: put the two systems together to relatively discuss, so the whole system has become complete, exactly the same stationary and moving rod systems as Einstein proposed. By contrasting the moving rod system and relatively stationary rod system, the answer is exactly the one given in Einstein's paper that when flying rabbit catches up to the turtle, the watches will begin pointing to different times, even if the flying rabbit and the dog are on different sides of the Pacific Ocean!

One question: how to "put" the two systems to relative?

Here is a fun problem.

In above fable, suppose the Flying Rabbit and the Tortoise is at each end of a long flying carpet, the rabbit ran to the other end toward the turtle with the speed of light. The turtle stay at the other end of the carpet, so the turtle has the same speed V of the flying carpet.

Q: Is this mode the same as the mode of the flying rabbit catching turtles? What are the speeds in this model? When the flying rabbit will catch up the tortoise? While you read the other articles about the theory of relativity movement, what kind of model is the system in that model?

Note that the following difficulties came:

If flying rabbit run then vacated landing legs, what is its speed?

Do not ask me, too complicated!

The reader can also discuss or even write a paper on such a situation, in which a wheel that runs as fast as the light, and rotates forward on the rod, and the friction between the wheel and the rod is perfect.

In sections §1 and § 2 of "On the Electrodynamics of Moving Bodies," with the comparison of stationary and moving rod systems, plus elementary arithmetic, plus lots of definitions, Einstein confused others, confused himself, and confused whole of humanity for about 100 years.

Always, someone has asked this kind of question: from the stationary system to see light in moving system ... and so, I say the simplest and crudest method is straightforward: if you separate the two systems into individual, independent systems without relative terms, which clock is subject to time disorder? When did it start to disorder? At which point did the confusion start? If there is no time disorder when each system is separated; but after did relative comparison of them in thinking, there is time disorder, the thinking causes the time disorder? And we only have one relative approach: that is, relative several systems in our mind. Otherwise, what can we do to the relative of the systems? Can they be tied up

Part I   A 4th grade math quiz that has been fooling human a century

relatively?

To make systems relative in one's "thought" can come up very funny scenario: I am sitting on an airplane. Suddenly, a guy in a car on the ground wants to relatively compare the movement of the aircraft to the movements of his car. Because of this comparison in someone's mind, my watch will be bewildered! Is it not funny?

If many vehicles from different directions each want to take a relative comparison with the airplane at the same time, my watch would ask me:

"Doctor, how do ye count the time?"

"Shut up! (我不知道. 闭嘴!) I'm really annoyed! Really, I don't know!"

How to relativity? the power of the thoughts!

Organizing the debris of time, tracing the wind from the flowing cloud. This is the human instinct and forward momentum.

In the study of the theory of "On the electrodynamics of moving bodies," I would like to re-simulate the idea of master, try to find out possible relationship between moving rod, static rod, and lights in real study, to explore how can effectively relative them to each other.

First of all I wanted to create a few completely independent systems, and then make them relative to each other.

Static rod is the easiest. Finding a rod, or at both ends of a distance make specified marks, and let it represent a static rod. Of course, this rod is completely self-independent.

Light moving back and forth at the ends of the rod, two mirrors can be placed at the ends of a static rod, and a laser beam A moves between the two mirrors back and forth. The mirrors and the rod are completely separated and may be very far apart and have nothing to do with each other.

And then in another place put moving rod.

The above three objects can be completely independent.

The laser beam B moving back and forth at both ends of the movable rod is more difficult to be independent from the moving rod. I designed like this: the two mirrors at a fixed speed along the straight forward, the distance between the mirrors is the rod length. Laser B moves back and forth between the mirrors. In this way, this beam of laser B is completely independent of the moving rod. It can be separated by only a few meters or a few light years apart.

These four objects that we constructed are completely independent from each other. Now we want to link them to "relative", how to do it? They cannot be tied together, or even cannot be gathered in a small space to facilitate research.

What we can do is to consider the static rod and laser A together to set up the static system; and consider the moving rod and laser B to combine into the moving system; and again the "consider" is in the mind: we consider of the static system and dynamic system "relative" up!

Please note that all systems and objects that being discussed above are grouped in terms of "consider" in my mind.

In people's THOUGHT, sub-group the real-world objects, systems, and relative them in mind, which can change these objects size, the system time, and their speed! Is this the power of God's thought? Einstein has been a scientific god for hundred years! Are we using relativity that is, using the power of God?

If you do not think "thought" is the only way to do systems relative, then how to do the systems relative in reality? What are the kind of relative rules or methods? Please advise.

I cannot imagine out!

Now we should know how to make spaceship's cabin like an accordion to elongate and shorten according to the theory of relativity: Johnny sits on the airplane, Doe in the high-speed rail, King in the car, Eugene is walking. Let them at different moments to imagine themselves in the movement relative to the same spacecraft. This relative result is: Johnny's movement relative to the spacecraft's cabin causes the biggest cabin volume contraction; second is caused by Doe; third by King, while the smallest contraction is caused by Eugene...

Wait a minute, write here suddenly having a thought: the cabin is shrink while relative to a moving object, then if it is no longer relative to any system, will the cabin elongation back? If it does not restore, then if many people continue relative to it, or if a person intermittently relative to it, will that spacecraft cabin be reduced to as long as a piece of paper thickness? If the cabin will restore while is not being relative, then what is the materials this cabin made from?

Also, why do we not see such a contraction?

And also...

Come on, you!

Fun, right?

The **THOUGHT** of Einstein's relativity is so powerful!

Part I   A 4th grade math quiz that has been fooling human a century

Conclusions

Everything in the world is in the movement, and never be absolute. The emperors dream to live ten thousand years; but billions of years are considered by Heaven and Earth as an instant. Relativity is the common sense people believe.

But Einstein's relativity principle is different. He defined the relativity principle, using it to prove relative concept; equaling the limit process and limit result; with rough mathematics same as the fourth grade movement superposition arithmetic brought human thinking into the confused trap for one hundred years.

Writing to here, I can't help but feel sad for the human: relativity principle is such a rough definition, unproven, self-contradictory, self-demonstration, how can it has been controlling human science thought a hundred years?

God's laughter, ah, has been echoing in my ears...

CHAPTER 2
GENERAL RELATIVITY THEORY QUESTIONED

Science should be always in progress, but not always stay inside the ancient giant projection.

According to Einstein's general theory of relativity, gravitational fields will slow down the speed of light, gravitational deflection of light occurs in the action, or the light in curved space-time deflection occur. For example, since the gravitational field of the Sun or its mass produces curved space-time, the rays that pass near the Sun will be slightly deflected. Figure 2.1 shows the figure in Einstein's manuscripts described the stellar light deflection by the gravitation field of the Sun.

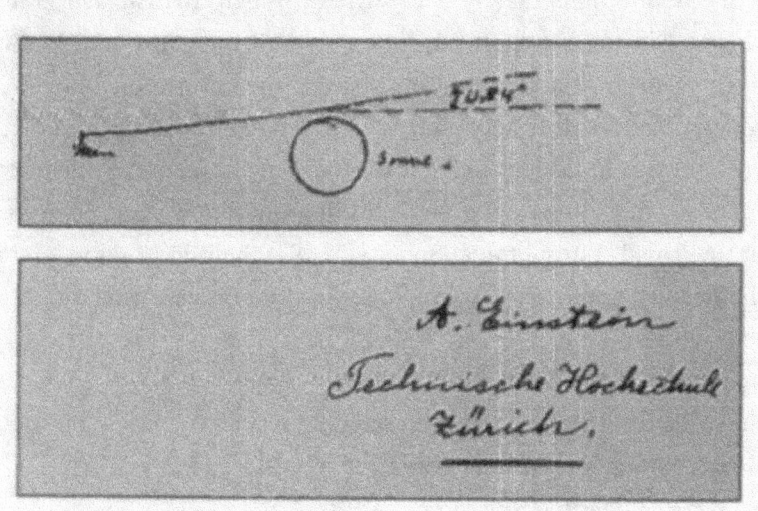

Figure 2.1 Einstein's manuscripts: Sun's gravitational field deflecting the stellar light

Part I   A 4th grade math quiz that has been fooling human a century

Of course, Einstein later developed a projections of a massive object's gravitational field near the curved space, which led to the theory of curved space-time theory.

However, whether it is his early theory that light will be bent by gravitational field, or whether a curved gravitational field will bend the light, every part of the theory, from the proposition, to test inspections, to practical applications, has questionable place, and exposed the irreconcilable contradiction between special relativity and general relativity theory.

In simple terms, there are two very intuitive grounds of objection:

First, imagine there is an opaque massless large carton ball with the same volume as the Sun. Will it bend the nearby passing light rays? Will this zero-gravity big ball form a "gravitational light cone" behind it?

Second, the star light fills the entire space, where can it turn?

We use these two reasons as the main clue to discuss the assertion that a gravitational field or curved space-time will deflect the light.

As usual, we first need a little background.

Observation history to proof gravitational field bends light rays

In 1915 Einstein suggested that a gravitational field would deflect light. Figure 2.1 was taken from Einstein's manuscript. Since World War I was under way, his predictions could not immediately be verified. When British astronomer Arthur Eddington heard Einstein's hypothesis, even though Germany and Britain were at war, he led his team to Sobral, Brazil to observe the eclipse on May 29, 1919 and saw the light deflection near the Sun, which proved Einstein's prediction that a gravitational field would deflect light. It was one of the most powerful evidence of Einstein's general theory of relativity. Eddington attended a famous luncheon banquet that the Royal Astronomical Society in 1919 held in London, and recited the following humorous poem announced that this important observation result that made Einstein world-famous:

Oh, leave the Wise our measures to collate,
One thing at least is certain, light has weight;
One thing is certain and the rest debate –
Light rays, when near the Sun, do not go straight!

Please note this verse: "One thing is certain, light has weight". If light has weight, it can be inferred that the speed of light must change, but this is against

Einstein's assertion in the theory of special relativity, which specified that regardless of source, the speed of any kind of light is the same.

Because light has weight, it can be bent by gravitational fields when the light passes a celestial object. From this viewpoint, there are two points to consider:

1. Do the celestial gravitational fields work on the rays emitted by the celestial objects?

2. Does the gravitational field of the Earth where the telescope is receiving the light have effect on the light?

Different-sized celestial objects have hugely different qualities, of course, and also huge differences among their gravitational fields. When the photons are being emitted, will the differences have any effect on those photons with different weight? Maybe people can barely argue that the greater the gravitational field of the star, the greater the power of the outbreak of the photon, two-phase offset, the speed of light cannot probably dramatic change?

Then, when receiving photons, the gravitational effect is an inevitable vulnerability that must be explained.

If light falls on the Earth's surface, of course, that light will inevitably be affected by the Earth's gravity that accelerates falling photons. And because above the Earth, light received by the Hubble Space Telescope, for example, is not bound by the similar gravitational field, the photons it receives will not be accelerated by the Earth's gravitational field, and therefore when we measure the light from the same celestial body in two different places (the Earth's surface and the Hubble Telescope), the results from the ground-based observations will be different from the results from those of the Hubble Space Telescope.

Here's something that requires more careful consideration: photons with different energy levels may have subtly different qualities. Although this difference is too small to be detected out by our current technology, after a remote transmission under the effect of the gravitational field or other forces, perhaps we can test it out.

Thus, in order to solve the contradiction of why the speed of light changes when the light is bent, the curved space-time of general relativity theory debuted.

Einstein's theory of curved space-time

Hawking wrote in his book "A Brief History of Time" (please note the following two paragraphs are references to the original, emphasis mine):

"The special theory of relativity was very successful in explaining that

Part I A 4th grade math quiz that has been fooling human a century

*the speed of light appears the same to all observers (as shown by the Michelson-Morley experiment) and in describing what happens when things move at speeds close to the speed of light. However, it was inconsistent with the Newtonian theory of gravity, which said that objects attracted each other with a force that depended on the distance between them. This meant that **if one moved one of the objects, the force on the other one would change instantaneously. Or in other words, gravitational effects should travel with infinite velocity**, instead of at or below the speed of light, as the special theory of relativity required. Einstein made a number of unsuccessful attempts between 1908 and 1914 **to find a theory of gravity that was consistent with special relativity**. Finally, in 1915, he proposed what we now call the general theory of relativity.*

Einstein made the revolutionary suggestion that gravity is not a force like other forces, but is a consequence of the fact that space-time is not flat, as had been previously assumed: it is curved, or "warped," by the distribution of mass and energy in it. Bodies like the Earth are not made to move on curved orbits by a force called gravity; instead, in a curved space they follow not a straight line but the nearest path, which is called a geodesic. A geodesic is the shortest (or longest) path between two nearby points."

First of all, since general relativity is "gravitational theory in harmony with special theory of relativity", we have proved the absurdity and erroneous of special relativity.

Then the description from Hawking *"Or in other words, gravitational effects should travel with infinite velocity"* is not correct. According to his own understanding that "this means that," "in other words," Hawking misinterpreted the meaning of Newton. We can say in such a way: This means that Newton describes a gravitational field or a gravitational ring or gravitational area that varies by distance around an object - although Newton did not use the term "field, zone," the simple gravitational formula he gives is applied to any point in the "area" or "field" surrounding the object. So in fact Newton defined a "field" with his gravitational formula without using word "field". You can not say that because Newton did not use the word "field", so his theory means "gravitational effects should travel with infinite velocity." Newton 's formula, or using Hawking' s words, means that there is a gravitational field around the object distributed in accordance with Newton 's gravitational formula! The gravitational force of a celestial body is not, of course, emit when another object approaches, instead the gravitation "always" is there! This gravitational force is represented by Newton's gravitational formula. The object enters the range where the gravitational forces that have permanently exists, the force act immediately. As the distance changes, the size of the gravitational force also changes. No any scientist would imagine, as Hawking thought, "Gravitational effects must be transmitted at infinite speed," because no one would think that "infinite speed" could be achieved, and also because gravitation does not need to be transmitted at all.

Gravity, according to Newton's formula, is just around the object. Whether you describe it with the word "field" or not, it is there, not too much not too few, exactly as Newton's law describes.

It's just like a light of the Sun or any lamp or any electromagnetic field: the space around it in a certain range is illuminated at all the time. Any object entering from the outside of the light range will immediately be illuminated, rather than objects come near the Sun then the Sun begins to shine and begins to transmit electromagnetic waves from it, and after some transmission time then the object becomes illuminated. With a change of distance, the strength of the gravitational field will change, just as the intensity of light changing from the light source results the closer get in more light.

Even in the Michelson-Morley experiment itself, something needs further explore. How far did the distance the light travel in this experiment? If light speed changes happen after light travels millions of light-years, is it possible for the Michelson-Morley experiment to simulate that? Even from the beginning of the Michelson - Morley experiment until now, the light beam in this experiment has been running all the time, the distance the light beam has walked through is still less than 200 light-years. How could people simulate any physical phenomenon that needs be calculated with units of millions of light-years?

Here I give a law: "star light speed will be gradually reduced after the light traveled a million years!" Please use experiment to prove that I was wrong!

Can you use the experiment to prove it? Can you mimic the light traveled even a thousand light-years distance?

Einstein uses light speed to describe his theory. The experts and masters of the physics want people who are opposing relativity to do experiments to prove the relativity theory is not reasonable. In fact, this is a bad kind of sophistry: the relativity theory can not be proved that it is correct or incorrect with doing experiments. Experiments like Michelson-Morley are basically not established! The detailed reason will be discussed in next chapter.

Part I A 4th grade math quiz that has been fooling human a century

Now, I will give two more laws, if you don't like them, please do experiments to proof that they are wrong:

I believe the photon has a weight, each photon's weight is one millionth of an atom.

I believe the gravitational field of a celestial body will deflect the light passing by it. This deflection degree can be calculated using Newton's law. In accordance with the weight of the photon I defined above to calculate, the degree of deflection can be negligible, so the celestial deflecting of the star light effect is basically caused by the diffraction properties of light. Thus Einstein's curved space-time field theory is completely wrong.

Above are my three assertions, please use experiments to prove that they are wrong!

Give you a hundred years to complete this experiment, is it enough?

According to Mr. Hawking, mankind has come close to fully understanding the universe: according to my understanding, human beings simply haven't touched the threshold of the real face of the universe.

Smaller human beings, not only can't across a span of a light-year distance of the solar system (but talking about hundreds of millions of light years of the universe model), but also never see the true face of the distant universe. (all distant images of the universe are false, which will be discussed in next part of the book.) I think that human beings should be aware of the limitation of their ability, abandon the arrogant of the "how bold people how high-yield," and calm down to rational use of taxpayers hard-earned money!

If every gravitational field would deflect light direction, a lot of funny results will happen: for example, the transmission of light through a curved path so that the light becomes longer, and thus from transmit to receive an average speed of light is slowed down. That is, the speed of light measured in different distance will be different.

On this point, Einstein explained with curved time-space, the shortest distance is a curved geodesic.

But the total route of the light emitting from the stars until received by the telescope, after is bent by many massive objects, the total distance the starlight passing has increased.

Even geodesic, which itself has a room that can be discussed. Let's look at Figure 2.2 below.

Figure 2.2 Is the Geodesic the shortest distance?

First, we have to abandon our fixed ideological that geodesic is the shortest journey between two points in a curved space like inside the ocean.

In Figure 2.2, note the distance between the two points A and B. If we are sailing in a boat, the shortest connection at sea level between the two points A and B is the arc that is a geodesic, because the boat must travel along the surface of the water. But if we use a submarine, is it necessary to travel in an arc between the two points? Especially if the submarine goes underwater, then is the shortest path an arc or a straight line?

In a vacuum, what will force light to move along a curve? Nothing!

So, Einstein's curved space is defined without first proving that light in a vacuum will be forced to go geodesic. Does Light in vacuum go straight or follow a curved path? Of course, it goes in a straight line. Then why does the space around the massive object become curved? Should it also be necessary to prove that the gravitational field will be bent into a curve by something? If the Sun is replaced by a cardboard ball of almost no mass and therefore no gravitational pull, will it bend the light? Of course, it will also bend the light. There is no discussion of this situation in curved space-time theory. So on a few mathematical formulas hastily put light bending phenomena into the definition on gravitational field caused curved time-space. This is not rigorous scientific reasoning.

Let us turn to the Figure 2.6 several pages ahead. That figure shows the proportion of the size of the Sun and light waves of its nearest stellar neighbor. Compared to the light waves of the nearest star, the size of the Sun is like the size of a sesame seed is to the Earth, so how can it bend such huge light waves with such small weight and gravitational field?

Part I A 4th grade math quiz that has been fooling human a century

THE PROPOSITION OF "GRAVITATIONAL FIELD DEFLECTS LIGHT" PROPOSITION MAKES ITSELF TO BE CRITICIZED

The light emitted by any celestial body is a 360-degree ball with virtually unlimited space to spread. The propagation of this light sphere size is light-years of age to the same number of light-emitting objects. For example, if a star is a billion years old, its propagation distance (calculated in accordance with the speed of light constant) is a billion light-years. It covers itself as the center of the sphere of a billion light-years light radius ball. Starting from the light-emitting celestial object, its light intensity is gradually weakened outward by distance.

Of course, as it spreads billions of light years this way, the light will pass through a myriad of large and small objects, however, there is no celestial object can change the propagation route of the light ball of the celestial. And when this celestial light filling the center of the sphere with the celestial bodies to a billion light years of space for the whole sphere radius, there may be a small part of the shadow of the back of each celestial body is blocked (it was discussed while solving Olbers' Paradox), all celestial light and space in this sphere have been the shining light ball inclusive. We can say that every angle is the propagation direction it proceeds to be, or it does not have any particular direction like a laser beam has.

How can one change direction if there is no direction?

If there is no dirt, where does dust come from?

So, from the perspective of wave propagation, the proposition of a "gravitational field that deflects light" is not convincing.

Figure 2.3 Sunlight fills the entire space   Figure 2.4 Sun's rays fill the entire space

From another perspective, if a celestial body changes the direction of the light propagation of another celestial body, then, what is the significance of astronomical observations? Because the distant stars we are observing today may be changed tomorrow by another moving star. Will our observation results have any consistency?

From the light-ray point of view that Einstein discussed only, the "gravitational field deflects light" also is not accurate.

Celestial light is spreading indefinitely in a 360-degree sphere in space. The quantity of these rays is huge. Suppose a small fraction of these rays pass by X celestial body. Some rays are absorbed by X, and another small part pass by X, but most of the light rays have no effect on it, and even spread away from X. Only a very small portion of the rays (approximately less than one billionth, depending on distance) will be affected by X. Does this small part represent the propagation direction of all celestial light rays? Does the gravitational field of X causes the deflection of light rays as Einstein drew in Figure 2.1?

Is this a few light rays that passing nearby X the Central Committee, representative of all the celestial lights launching to the 360-degree space?

In addition, the distant starlight ray is received from the transmitter to the telescope, on its route how much of the gravitational field of celestial bodies does it go through?

If gravitational fields make light bend, so in Figure 2.5 in our night sky what kind of picture would it be? Just the thought of almost every distant star light ray being bent by countless huge, constantly moving gravitational fields (such as black holes, superstars, etc.) and continuously twisted, why there is not even one star light that follows the twisted fields dancing? The night sky that has been playing upon safe serenade for thousands of years without crazy dancing, is it calmest endless direct protests upon Einstein's bending starlight argument?

Figure 2.5 this is a photo of suspected high-redshift galaxies released by NASA. Toward Earth so dense stars, why no star is dancing under the baton of Einstein's gravitational?

So, even if just from the point of view of light propagation, the proposition of

Part I A 4th grade math quiz that has been fooling human a century

the "gravitational field deflects starlight" is open to criticism.

It can probably be answered thus: during the starlight propagation process, when it encounters another celestial body X, since the starlight is blocked by X, the back of X will form a shadow of same large size like X's. However, a small fraction of photons or light waves in the near vicinity of X will be affected by some action, so these light rays that are near but are not blocked by X, will gradually bend into the shadows. This bending process is gradual, thus the partially shaded light will gradually erode in space and finally form a local, tapered shadow space, the so-called "gravitational lensing" (I think the term "dark cone" is more appropriate because if there are no gravity obstacles it will still produce the same shadow space). But this will have no effect on the continuing spread of the starlight.

So, how big will this local disturbance bending be? Exactly what factors caused it?

Einstein believed that this is entirely due to curved time-space or the gravitational field of a huge celestial object. However, careful calculations will call this conclusion into question.

The Stars closest to the Sun is Proxima Centauri, 4.22 light-years from the Earth. The ratio of its radius to the Sun's is approximately 1:56,000,000. How big is this ratio? Suppose a man is one meter height, the size of haze particles is 1 micron, the proportion of haze particles with human height is 1: 1000000. Please carefully compare above two ratio numbers.

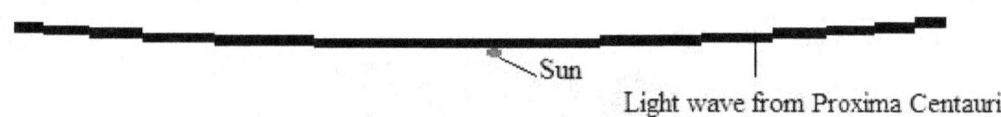

Figure 2.6 Schematic diagram of the ratio of the light wave from Proxima Centauri compared to the radius of the Sun. If drawn in actual proportion the Sun would be much smaller.

Therefore, near the celestial body X a small portion of photons passing through may be affected by the gravitational field or curved space-time of X, but this effect is extremely small, only a local perturbation that does not affect the overall spread of celestial light, with limited affection on a small part of photons. Einstein's relativity theory should clearly indicate this.

Figure 2.7 Spindly sandpiper legs have no impact on an ocean wave

Due to the relatively small objects unlikely to have a huge impact on the relatively huge super big light wave, we need to ask: are there any other factors at work so that the partial light fold the shadow cone formed lenticular space? What is the quantitative relationship between the dark cone shadow space and factors such as gravity and size of the celestial body? Can we possibly to calculate the size of this dark cone shadow space by just entering certain data related to the celestial body?

According to Einstein's general relativity theory, the light deflection angle Θ of gravitational lens can be calculated by the following formula. The angle of deflection is

$$\theta = \frac{4GM}{rc^2}$$

Toward the mass M at a distance r from the affected radiation, where G is the universal constant of gravitation and c is the speed of light in a vacuum, regardless of the size of the object causing the deflection.

The above is the formula for calculating the "gravitational lensing" based on Einstein's relativity theory. It basically rules out the influence of other factors in addition to the gravitational field, and doesn't have much relationship with the size of the celestial that cause the deflection.

However, if the Sun is replaced with the same size but without gravity cardboard sun, what is the shadows it formed?

Our purpose in designing the cardboard-ball experiment is that through experiments to find the relationships among shadow cone, cardboard size, distance, and various factors interacted, so that not only can we calculate the deflection of light and give quantitative description that the Sun as a pure obstacle cardboard Sun, but through further calculation we can arrive at the light deflection effect size

Part I   A 4th grade math quiz that has been fooling human a century

caused by the curved time-space gravitational field.

FINDING ALL FACTORS WHICH ARE BENDING THE LIGHT THAT IS GOING THROUGH NEAR THE SUN

Thanks to the cardboard Sun experiment we already know, in addition to gravitational field (or curved space-time), that diffraction also deflects nearby stars light, so a small portion of photons in the vicinity near the Sun will be affected by the Sun, resulting in a limited short-term local disturbance. Whether this effect is generated due to the gravitational field or due to diffraction effect, or due to both, is important to determine. Further, if it is a common action that generates the light cone shadow, then how much light is bent by gravitational field, and how much is bent by diffraction?

This must to be identified by experiments to quantify. We cannot only talk about gravity or curved space-time, while totally ignoring the other important factor: the diffraction of light. This became especially important after Einstein changed his gravity field theory to include curved time-space. He did not use gravity, so photons became a particular matter without weight. However, the light diffraction phenomenon does not require gravity either. How can we ignore such an important factor?

Consider this: if we replace the Sun with a cardboard replica of the same size, does the cardboard sun block the light going through it? Will this sun form a large shadow behind itself? Does the light need bend to fill the shaded space? What causes this light bend? The cardboard sun has no gravitational field! A relatively big light wave, even without any external factors, will form a shadow cone space behind the cardboard sun. We name this shadow space a shadow cone. This shadow cone is produced by diffraction without the slightest gravitational or curved space.

So: the bending effect of celestial light through local disturbance should be the gravitational field or the bending space and the diffraction, even including the results of the propagation properties of light itself. In all factors, the celestial scale itself is the most important factor; followed by then nature of light diffraction; and the gravitational field of the light deflection effect is very small, even no way to measure at all. Such speculation is based on Newton's gravitational formula. The photon quality tends to be infinitely small, so the effect of the gravitational field tends to be extremely small as well.

As can be seen from Figure 2.8, either the gravitational field (or curved space-time field) and light diffraction phenomenon are locally bending the light from the celestial body, to fill the dark space caused by the Sun. These two factors are likely to coexist, and where the main work is the diffraction properties of light waves. The phenomenon has been described in my book "《谁有权谈论宇宙》 (Who Should Talk About Cosmos") published in 2005. Ten years later, I put forward the relevant observations of experiments to confirm this speculation, and did a couple of extremely rough preliminary experiments. Here's one of them.

The purpose of the cardboard experimental design is to finally get a quantitative result that can be determined after the oncoming waves of celestial bodies are blocked out by the Sun as an obstacle, and through this experiment we can find the basic quantitative law from the following: the unobstructed light continues to spread across the Sun, how much the gravitational field (or curved space-time) makes light bend, and how much these light waves are bent by diffraction property.

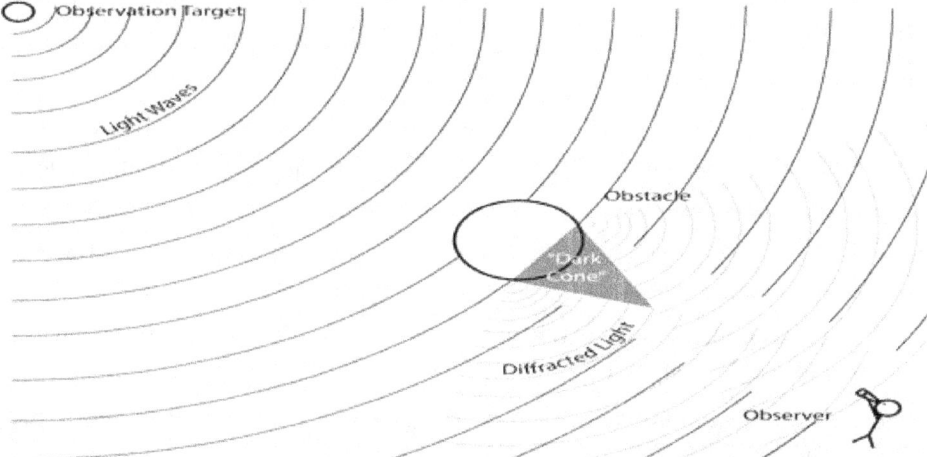

Figure 2.8 A dark cone phenomenon will be generated after the celestial light goes through the sun. This phenomenon was caused not only possible by the Sun's gravitational field or curved space-time, but more likely caused by a barrier obstruction that blocks away the light because of the impact of wave diffraction, and the gravitational field impact is very small.

Because of our very limited methods of measurement, we cannot fully appreciate the true nature of light. Ours is a very superficial understanding. Suppose the weight of the photon is a thousandth of a millionth of an elementary particle. Do we have any method to measure it accurately? If not, is it reasonable to say the photon has no weight?

What we need to think about is: Imagine replacing the Sun with a similarly-sized cardboard ball that has no gravitational field. How much light deflection will happen after this light passes over the cardboard ball? Then observe the real Sun, and see how the occurrence of the different deflection of light changes.

Part I   A 4th grade math quiz that has been fooling human a century

This changed different portion is produced by the gravitational field. We expect this change will be small, and will be difficult to measure.

So it is particularly important to calculate the diffraction deflected by the light.

Looking for all factors that affect light bent - preliminary cardboard experiments

We have done some preliminary experiments to test if diffraction can deflect light. The rough preliminary experiments can prove such an idea: If the sun were made of cardboard, it would certainly have no gravitational field, and would also be impossible to produce curved space-time, but this cardboard sun makes light waves in its bend by nature due to diffraction. When we can quantitatively calculate the diffraction effect, then according to the observed solar eclipse shadow sizes on the Earth, we can calculate how much the light is deflected by the gravitational field or curved space-time.

Figure 2.9 below describes our cardboard experiments to prove that under certain conditions, the cardboard with no gravitational field also can deflect light and produce a light dark cone similar to the gravitational cone. Ten years ago, in my book, I depicted the phenomenon that an aircraft will project as different-sized shadows on the cloud vs. on the ground, which is also a similar proof.

This experiment is only of the most preliminary kind. Using different sources, at different distances, with a piece of cardboard of different shapes or different sizes, we can draw relevant diffraction formula, then use it to calculate the diffraction effect for the Sun blocking the stellar light from different light sources, thus separating the gravitational field effect and diffraction effect.

DESIGN THE EXPERIMENT TO LOOK FOR ALL FACTORS THAT AFFECT LIGHT BENDING

Figure 2.10 below is the expression of our initial ideas and simple mathematical model of our test.

The purpose of this test is to calculate the width of the diffraction annular R_i.

In Figure 2.10:

　　LS is the Light Source),

　　CB is the Cardboard),

　　PS is the Projection Surface,

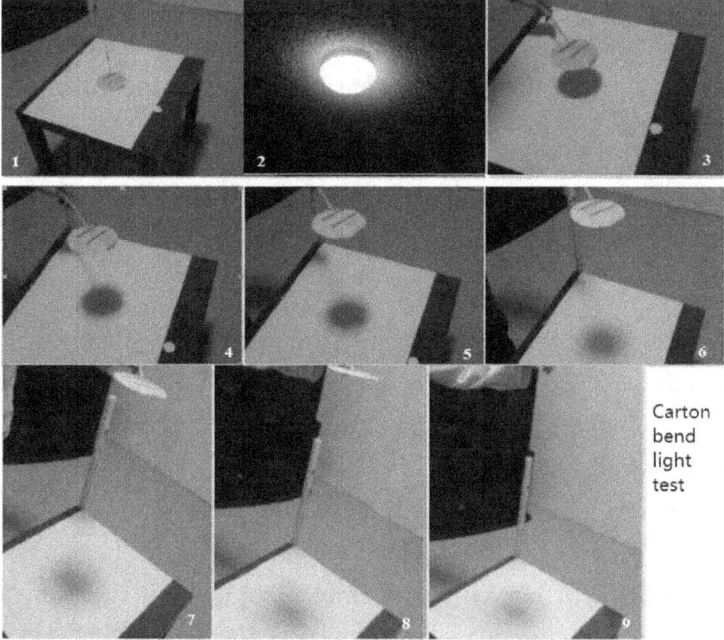

Carton bend light test

Figure 2.9 The whole picture is divided into 9 thumbnail frames: frame 1 is the light projection surface, the distance between circular pieces of paper to the surface will be gradually increased; frame 2 is a projection surface directly above the light source at about two meters. With this shape shade light disseminated quickly, otherwise two meters of distance cannot produce the same experimental results; frame 3 begins to move the cardboard circle a little upward, you can see the shadow of a very neat circle projection plane; frame 4 to frame 9 is the round cardboard is gradually moved upward toward close to the light source, we can see with cardboard moving near the light source, the surrounding light shadow of the circle on the projection surface is increasingly eroded, which is totally due to the light diffraction effects. Round cardboard does not have a gravitational field.

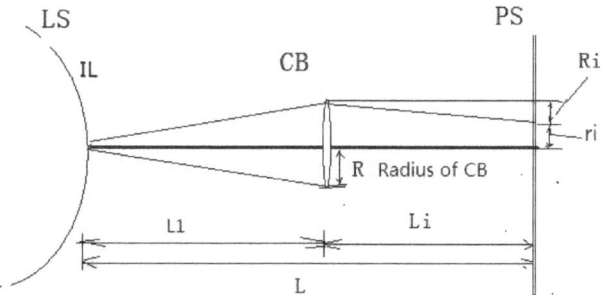

Figure 2.10 A test to determine the diffraction ring size of the circle of cardboard blocking light generated in the projection plane. In this experiment, you can move CB as shown in Figure 2.9, you can also fix L1, change Li only, or you can change both L1 and Li.

Part I A 4th grade math quiz that has been fooling human a century

R is the Radius of the cardboard),

r_i is the shadow radius on the projection surface, the distance from this point corresponding to the projection surface of the cardboard circle is L_i,

L_i is the distance between CB and PS

R_i is the light ring width of the diffraction ring width around the projected shadow

$R_i = R - r_i$

With the change of Li, r and R are changed, in general, should have:

$r_i = R - R_i$

The r_i can be obtained by measuring the shadow of the projection surface.

$R_i = f(LS, L, L_i, R)$

Finally, the results can be used to calculate the Einstein light deflection phenomena. As shown Figure 2.11 below, the eclipse due to the additional effect of the gravitational field (or curved space-time) of the moon, the Sun rays not only have similar deflection caused by the cardboard diffraction, but also the effect of curved space-time or gravitational field. Assuming all the aura width of light deflection is Bi, we can set up a testing model follows.

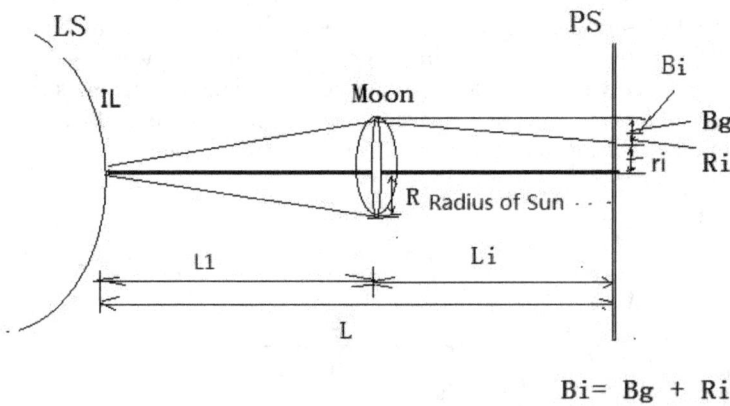

Bi = Bg + Ri

Figure 2.11 experimental model to distinguish the moon light deflection factors: Bi is full moon light deflection aura wide, it has two effects Bg and R_i aura, of which Bg is caused by the gravitational field or curved space-time, and R_i is caused by the diffraction.

Let

Bi be the aura width of all deflected rays of the Sun

Bg be the halo width of the Sun rays deflected by gravitational field or curved space-time

R_i be the light ring width of the diffraction around the projected shadow

r_i be the radius of the moon projection, obtained by field measurements

R be the diameter of the moon

$R_i = f(LS, L, L_i, R)$

We can get:

$B_i = R - r_i$

$B_g = B_i - R_i$

Thus, we can distinguish the basic amount from the gravitational field (or in curved space-time) and diffraction properties of two different factors that make the light bend.

Unfortunately, I just an amateur doing a little test, with many variables. I can't complete the required experiments that might quantitatively rewrite the theory of relativity.

Conclusion

Ten years ago, in my book "Who Should Talk about Cosmos", I described and compared the aircraft projection on the cloud and on the ground, which is the seeds of thinking in this chapter. With any opaque object to block the light from the Sun or from a lamp, in the back of the object will form a shadow cone. This shadow cone has nothing to do with gravitational attraction, and is entirely the result of the diffraction effect of light. Such a normal physical phenomena, How could Mr. Einstein make it magic-like confused disappear in the universe, leaving only a distorted space-time?

Several sections of the above discussion illustrates how, over the past 100 years, combined with modern technology and new achievements, we reconsider the conclusions of one hundred years ago to reflect and promote the continuous progress of human technology.

Part I   A 4th grade math quiz that has been fooling human a century

CHAPTER 3

DISCUSSION ABOUT TIME AND SPACE CONCEPTS

SPACE IS A PHYSICAL OBJECTIVE REALITY

Space and time are essentially different.

Space is a physical objective reality that is "empty". This presence is visible, accessible.

We tell people, no matter is human or an alien from far away planet: The empty cubic location between the Sun and the Earth is an example of "space", these words everyone could understand. We say that above the playground is space, let's go there to run. Everyone knows what this means.

Space is particularly important. If there is no "time" concept, everything will still alive. But we simply cannot live without "space". Imagine if we exist in a place where there is no space (of course, there never was such a place), how terrible scene it would be.

There is no need to make a special definition of space, since it exists by itself, same as the existence of other entities, although it is "empty".

We can, merely for convenience or mutual exchange, come up with a number of related terms. But it is not necessarily, only for humans to communicate and description. For example most people around the world use the metric system to measure space, but the United States has insisted on the imperial measurement system and it works just as well. We can also say the distance between two celestial is "100 times of maximum distance of the Earth to the Sun", or say the amount of land size the "thirty paces wide," and so on.

In fact, it is according to the size of the different space that humans identify the celestial objects as stars, galaxies, clusters of galaxies, and so on.

Part I A 4th grade math quiz that has been fooling human a century

Space is no need of other kind of assistance in explaining itself.

Space cannot be arbitrarily changed.

Because space is an objective reality, it cannot be as free to change as "time". Suppose the Sun and the Earth appear to be at a fixed distance to the people on Earth. This distance will not change because of other objects moving with a speed close to the speed of light. When the Sun and the Earth do not have any change, it does not change the distance and the space between.

Imagine a person is encased in an iron box space tightly oppressed him. If the space in the box is magically reduced to smaller down again, he cannot continue to survive. Space cannot be arbitrarily changed, nor can change by the "time". The so-called "time space exchanges" here is completely not true.

People use a variety of definitions, various models, and variety of coordinate systems to describe the space, which are reasonable. The one that is generally accepted, may become the most commonly used.

No matter what kind of description, space is there, not born immortal, do not increase, can't be changed by the description of people. The so-called space-time conversion, is the game that people talking about, is a description of human own feelings, has never happened in reality.

TIME IS NOT AN ENTITY, BUT A CONCEPT DEFINED BY HUMAN ON THE EARTH, A RULER

If an alien landed on Earth, at first it might not understand our concepts of years, months, and days, because they are based on the relationship between the Earth and the Sun. According to the relative motion of the Earth and the Sun the native people on the Earth defined them. Full revolution of the Earth around the Sun is a year, the Earth's own revolution is the day, and then divided into aliquots of uniform hours, minutes, seconds, and so on. This newcomer to Earth has not got inside this relationship and would not be familiar with this system. Saying "year" or "light-year" to an alien, he will not know what we're talking about if he didn't do a research before his arrive.

Objects working in the universe, operations of the universe itself, do not need the "time" bundle. No "time", any object movement in the universe will not be the slightest change. Even early humans did not need the "time" concept, but still manage to live well.

We cannot visually see "time", it only exists in our minds, and it is only a concept, a human-specified ruler with a single-line direction, a tool for describing the development and progression of operations of objects.

Only the evolved intelligent creatures in the universe need to use the concept to communicate with each other, they do need the concept of "time". When people want to express the passing of history, and describe expectations of the future, they need a way to explain it. Therefore, "time" is a "concept" rather than real object! We do not see a specific "time" object. "Time" is not essential for the operations of the universe, but the concept used by the evolved intelligent beings of the universe when they need to use the concept. They use the "time" concept to express or record events sequentially happened in the universe.

Time is a concept defined by humans on Earth. The traditional definition of time is: a description of the order of events happened. The fact that this order cannot be reversed or upset is no doubt. This is the "time" definition generally accepted by ordinary people. Almost all mankind uses this definition of "time" concept meaning to communicate. They say a simple "time" to each other and can easily understand what the other was saying. E.g.:

Question: What time is it?

Answer: It's 12 PM.

This simple answer contains some of the connotation components the concept of "time" definition, such as "PM," is the day divided into 24 equal parts of the unit, also implies what is the day, what is the content of the year. If the answer is "2016 Thursday, January 7," it would be more exciting, the short answer which even contains a long historical background of human history. It also implies the days after the Sun has passed over the highest point to the Earth in the sky...

"Time" is the definition of the concept of the connotation of the Earth, such as years, months, days, etc., or as a recognized and used to define the scale. With the evolution of science, the time scale has been getting more and more precise. Conversely, if thousands of years ago, said in 0.001 seconds, nobody knows what that were saying, there is no need to use such a short time at that years.

This time scale has an important property: if the time scale is changed, then all time-related events have subsequently changed in proportion to the scale of this time use within the entire system, without exception. For example, China's Beijing time is different in 16 hours from US EST New York's time system in the non-summer time (under Daylight Saving Time), then the time of all of China's clocks, watches, etc., are set in accordance to GMT. The time to change is the system change, everything within the system must comply with the provision.

It is very important to identify it is the systematic change, or an individual object change affected only by particular system factors (such as high-speed motion system). We have used such concept a lot in previous chapters.

People on Earth have a common identity on the definition of "time" concept, and this definition has a strict meaning that cannot be changed. Time as a ruler, measures the sequential evenly progress of events, and this is the most fundamental basis of people on the planet to communicate.

If someone wants to modify the definition of time, to turn it into something more mysterious or dazzling, he is self-modified the definition of the time, and before this definition is universally accepted and is used for the communication of mankind, he has to keep it and play with by himself.

Like I want to define a 100 dimension time, no problem, I can do it! Well this is a free world. If I have the authority, or the brain of someone magazine editor is in confuse, or I have this no one understand definition written in my book and is published, still it does not mean that this absurd definition of 100 dimensional time is reasonable, and the public would use my 100 dimension time to communicate, the world would run according to my definition, and my definition would be reasonable.

So, if someone want to put time plus a variety of arrows, or ended up in high-dimensional time, then printed in a book published, I can only say that he gives his own definition of time. Not means that his definition is accepted by the public, not therefore could prove that he could change the existence on numerous historical events to really exist at any or all the time, so that he could time travel back and forth in and out of those historical events.

Time Travel is simply an ideological farce.

We envisage time as an "absolute background." It exists everywhere, in all things, but never expresses itself. It is the only, omnipresent. It appears when you need it; it is hidden when you do not need it. It can be said that it is such a background material, it can be said that it is only a ubiquitous everywhere available concept.

The "speed that is close to the light speed" is of no use concept

This section only relates to the transmission of light in the vacuum of the universe, nothing to do with the light travels in different medias.

Because according to Einstein the substance would be converted into energy at the speed of light, but the concept of relativistic arguments are mostly discussed in events at the speed of light, so that "only a few experts" who understood of relativity, came up with the "close to the light speed".

But it is really an ugly concept.

All the light, electromagnetic waves, etc., are moving with the speed of light, so there is no such speed that near the speed of light. According to Einstein's theory of relativity, the speed of light is an unchangeable constant. In recent years, when it was reported that, the speed of light observed was different, or changed the speed of light by experiments, were not accepted by mainstream scientific community, and disappeared. Therefore, the mainstream scientific community does not agree with the speed of light can be changed.

Then can we develop a spacecraft that travels near the speed of light?

No!

First, humans in the foreseeable future cannot build something like that.

Second, even if humans made such a spaceship, it would be useless!

Why?

Because before the "Moving meter stick is shorter" paradox discussed in previous chapter is completely abandoned, the spacecraft is likely in danger of disintegration as it approaches the near speed of light. Even 10% of the speed of light, there is a risk of this.

Furthermore, we can estimate: what more reasonable near-light speed should be? Or what really is "close to the speed of light"? 80%, 70% or 60% of the speed of light?

How long can a spaceship flying at 80% the speed of light maintain contact with the Earth?

Even at 60% the speed of light, in just one year or two, communication between the spacecraft and the Earth will require a few years to be transmitted back and forth, and after a little further, about a few years later, communication lines will be completely lost.

That is, even created a "near the speed of light" spacecraft, the ultimate result is bound to lose all contacts between the spacecraft and the Earth in just a few years later, no exceptions.

Neither the universe having any object or radio waves moving at "near the speed of light", (according to Einstein's theory of relativity, the speed of light is constant), nor artificial made and no need to create "near the speed of light" spacecraft. So talking about "near the speed of light" basically does not make sense.

Therefore, "near the speed of light" is a statement with no practical value and

Part I A 4th grade math quiz that has been fooling human a century

confuse the concepts.

"Close to the speed of light" concept, there is an extremely harmful role, that is, to prevent people from fully understanding the theory of relativity.

Because according to the relativistic formula to do the application computing, needs to be in high speed that human can't make to be able to discern the results. But the speed human may reach is too slow, the results obtained can only be infinity close to Newton's classical laws of physics. That is, the use of Newton's law.

For example, the spacecraft cabin elongation and contracting quiz, we can't get the results with the experiment, because the speed of the aircraft is too small, so the calculated results by human fundamental measurement cannot be detected even the relativity theory were correct. Other experiments about relativity, most of them are the same cases. Even on the spacecraft or the satellite or any fastest object human can handle, its speed compare to the speed of light is not worth mentioning, cannot let people get good measure results!

In the fastest man-made spacecraft to do the "moving meter stick is shorter" experiment, since the spacecraft speed is 800 meters per second, is about 1/35,000 of the speed of light, so even put a rod as long as the length of the spacecraft, say 100 meters, calculating according to the Lorentz transformation, the shortened length will be about 0.0003 meters, about 0.0003% of the original rod length. No instrument can separate such a minor change from all kinds of disturbance factors.

This book has designed four experiments. Originally thought was that the in-depth discussion of the theory of relativity will be based on them. Now I found that these ideas are too naive. Three of the four experimental designs for testing relativity theory could not yield useful results. Only the last one that was designed for testing the redshift theory is able to succeed.

This situation is caused, no matter how the design, the results of the relativity experiments, are always difficult to achieve more accurate than the error specified by the instrument. That is to say there is no way to effectively test the theory of relativity. However, according to the concept of "close to the speed of light", the theory of relativity has its own space, also seemingly people have been doing a variety of experiments. So although a hundred years there are people has been questioned the theory of relativity, but it is always difficult to test from the experimental results. We can't prove it right by tests, and we can't prove it wrong by tests. It is such an absurd situation. Thus, the ideological argument is even more important.

Using reasonable scientific research method - Criticism on "Conversion between time and space"

On the same issue, there can be a variety of views, and a variety of research approaches. However, there should generally be a relatively good method. Generally authoritative approach is always the best way, but it is not necessarily absolute true.

Modern cosmologists use the concept of "time and space" extensively. What is the advantage of mixing using time and space, the two totally distinct concepts? It actually involves some basic principles of science and philosophy. They can be divided into the following aspects to consider:

First, the necessity to consider the scientific definition;

Second, the application point of view;

Third, is it a scientific action to replace the old concept with new one? What are the advantages and disadvantages of such replacement?

From the scientific definition of necessity after careful consideration, we find that mixing "space" and "time" for the purpose of research, not only does not have any benefit, but does have a lot of harm to cosmic research. Using space and time separately can clearly distinguish and describe events, but mixing time and space even can't draw out a four-dimensional space-time diagram, and there is no way to quantitatively study the details of cosmological events. We can draw a time axis of the development of events, and we can also draw out spatial axes, but who has ever used or seen something in the study of events, with "time-space" as an axis drawn as a quantitative image and can clearly describe the time or space?

From the point of view of applicability, mixing "time" and "space" together to discuss related issues will generate a lot of unnecessary trouble. Basically, it is limited only to qualitatively discuss issues; it can't be used to discuss quantitative details. For example, the Minkowski space-time diagram of two-dimensional space to be discussed in the next chapter, only discusses an event moving in a particular fixed speed. He used V / C as the basic unit of measure. Once the velocity V vary, he had to have additional drawing to go on. In this way, how can we do research on the various changes of V? From the foregoing discussion in previous chapter of Einstein's theory of relativity in his groundbreaking paper, we have seen how he mixed "time" and "space" with speed to confuse issues and concepts, confused himself, and whole world.

Finally, mixing time and space in discussion is not a good scientific method. In science, the most important thing is to clearly separate the factors that affect the

Part I A 4th grade math quiz that has been fooling human a century

object of study to figure out the impact of each factor on the research goals each from all aspects (quantity, quality, relationship, etc.) so that we can clearly see the effects of all kinds of objects in our study.

The meaning of the word "science", like the Chinese ancient characters "格物" or "科学", is classification "the sort of things." Science is the classification of knowledge.

First we must separate things to understand. Then integrate them together again to study.

The nature of space and time are two totally different object and concept.

Space is an indispensable objective physical presence. Humanity cannot survive without "space"; but people can live perfectly well with no concept of time. They are measuring completely different events. But because for any cosmic object, if you want to describe its movement in the universe, then the object is inseparable from the existence of this space and time. Therefore we can say the characters of the law of space in which objects exist is totally different from that of the time that record the regular sequential order of event development. The performance of the space and the time are completely different, but also inseparable and rely on each other in the usage of intelligence community as two basic units of measurement.

Einstein was able to take advantage of the speed of movement, and systematic observation from different subjective view, combined with dynamic calculations, **exchange conceptual time and physical space** into each other. It can be said that it is beyond the conventional imagination. No wonder only a few people in the world can understand it.

From the contrast of formula (5) and (6) in previous chapter, we see an instance that Einstein convert the speed with space in (5). The speed of light was a half while the light traveled space length was cut in half. Perhaps this is the secret of time and space conversion - convert between time and space through the middle concept of velocity?

Einstein repeatedly stressed in inertial system that no one can have a special status. But in fact when he was discussing internally within an inertial motion from another inertial system, he put himself in a special position, otherwise the two unrelated high-speed movement of the inertial frame, how can people in one internal moving system be directly researched by people in another system?

Back to the flying rabbits, turtles and dogs in the fable of previous chapter. How could the dog on the playground in California know the details of the race on the track in China? How it can affect the watches each of them is wearing?

Another point: we have always firmly stood on the Earth to look at the sky. Even on a satellite or spacecraft, from the big structure, it can also be taken as if on Earth. We never have the opportunity to board a spacecraft that moves "near the speed of light" to see the movement inside another high-speed spacecraft. We have almost no presence of man-made or natural "near the speed of light" motion (constant speed of light like Einstein insists) of the flying objects.

Thus, there is no solid scientific basis to support the method of mixing time and space together or transforming one into the other.

Break down the layered complex object systems

In scientific research, the system being decomposed into component-level subsystems is commonly used method. But sometimes it is strange that people make a problem that can be easily solved by decomposing it into subsystems more complicated. Then in the course of this man made complication introduce in some strange concepts.

Let's look an example from "Brief History of Time".

"The lack of an absolute standard of rest meant that one could not determine whether two events that took place at different times occurred in the same position in space. For example, suppose our Ping-Pong ball on the train bounces straight up and down, hitting the table twice on the same spot one second apart. To someone on the track, the two bounces would seem to take place about a hundred meters apart, because the train would have traveled that far down the track between the bounces. The nonexistence of absolute rest therefore meant that one could not give an event an absolute position in space, as Aristotle had believed. The positions of events and the distances between them would be different for a person on the train and one on the track, and there would be no reason to prefer one person's position to the others'. "

Here are a few important issues to be discussed.

First of all, in scientific research in accordance with our general approach, I will take the train and the man standing on the tracks as two research objects and put them each into one research system, then take objects inside the train as a research subsystem. Thus, there will not occur any conceptual confusion.

From our engineering science person point of view, from a hierarchical or a complex system that can be broken down into subsystems, to leapfrog directly into studying details of another level is simply very unscientific behavior. And to study

Part I   A 4th grade math quiz that has been fooling human a century

directly from one still object inside a still system to do research on a internal object of a moving subsystems, is very strange behavior.

For example, if we want to design a house, first we design the house shape, frame, etc. But we will not work on the interior design at the same time we're still calculating the amount of stress the girders can take. In fact, it is entirely possible that another team will tackle that.

Flight control center concern is the overall state of the aircraft, and will not pay attention to the fan movement inside the aircraft, we'll leave that to the aircraft pilots to study, but not directly from the control tower. In this way, by using system decomposition and integration methods, we may be able to describe aircraft interiors fan movement and the movement of the aircraft, even if the aircraft is doing complicated flight maneuvers. To directly study the movement of the running fan inside a flying aircraft from the control tower sounds very complicated.

In addition, from the perspective of system analysis, how can the man standing on the tracks know the activities of the little ball inside the running train? In fact, this is because the man who raised this issue is standing on a "superior situation than others" status.

Without this man who is in a "superior situation than others" status, the man standing on the tracks will not know, and will not go directly to research the table tennis in the train. Like there are two trains, each will not know or care about any inside motion of another.

Two spacecraft moving with the speed near the light will not care about the details within the other spacecraft. Headquarters will keep surveillance on the ship movements in space, but will not directly measure the clock situation inside the other spacecraft. The pilots will be asked to do that. By directly controlling the motion and communication of the spacecraft, it is easy to make the final overall conclusions by using the observed data from the pilot.

In the fable of the flying rabbit, turtle and dog, the dog running back and forth in California of course will not know the situation of the motion details of flying rabbit and tortoise in the other side of Pacific.

The complex system is decomposed into subsystems according to different situations or different levels of the system, the problems are solved one by one in the subsystems, and then the results are combined to draw final conclusions about the overall system. This is the general method of scientific research and engineering applications.

Do not complicate things, then draw conclusions from the complexity.

We can directly use a four-dimensional space-time event diagram that will be discussed in next chapter to discuss the previous examples of the train, table tennis, and the person standing on the tracks. The diagram can express the motions and the positions of all objects in the example. Due to the person who playing table tennis and the table tennis are together on the train, both of their movement and position in the world line will be completely overlap.

In the world line diagram the man standing on the rail will be a point, his movement and position unchanged. Thus, first the expression of system decomposition, and then merge what we are concerned to see the changes, our final conclusions won't appear weird. We follow this method in our daily work and life.

Part I   A 4th grade math quiz that has been fooling human a century

CHAPTER 4

THE EXPRESSION AND APPLICATION OF FOUR-DIMENSIONAL SPACE-TIME DIAGRAM for the World Line of Events of an Object

Minkowski Two- or Three-Dimensional Space-time Diagram Can't Clearly Describe Events Occurring Sequentially in Space

This chapter is not translated from Chinese. It may be possible to read this chapter about 10 lines at a glance.

In 1908, Minkowski developed his space-time diagram, providing "an illustration of the properties of space and time in the special theory of relativity." The purpose was to provide a tool for understanding relativity concepts like time dilation, and the shortening of a moving stick. This is called the Minkowski Diagram, or Time-space Diagram, also known simply as a space-time diagram (Figure 4.1). It has been used ever since. What follows is the simplest Minkowski Diagram. From it, we know that basically it is an imagined representation for the Lorentz Transformation.

A Minkowski space-time diagram provides "an illustration of the properties of space and time in the special theory of relativity." This is accomplished by curtailing space into a single dimension, allowing space-time to be expressed in a two-dimensional diagram. This diagram reduces space into one dimension, using the other axis for time. Often, the diagrams are arcane, and few possess the ability to comprehend their meaning. Even the light cone curtails space into only two dimensions, expressing space-time in a three-dimensional diagram. This limits the

Part I   A 4th grade math quiz that has been fooling human a century

ability for users to comprehensively describe even the simplest events occurring in space.

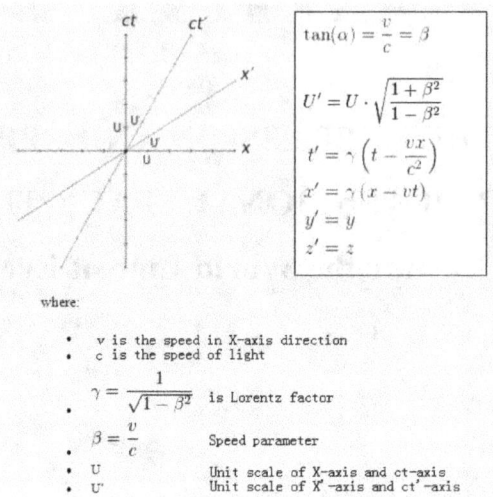

Figure 4.1 Minkowski's 2-dimension diagram and Lorentz transformation.

More seriously, this space-time diagram is unable to express the relationship between different events of objects. For some objects, event developments that should have heritage lack coherent expression. It makes extensive use of speed, but is limited in its use of distance. And the velocity v gives a fuzzy sense of space and time, so that the two completely different physical quantities can be converted to each other, as we already discussed in the previous chapter with Einstein's original relativity paper.

All of these cause defects, because the Minkowski space-time diagram lacks the visual images to express four-dimensional space and has to be based on a two-dimensional image. In this two-dimensional image the time axis is the vertical axis, and space is represented with the horizontal axis.

Since only one-dimensional representation of spatial distance or spatial axis is possible, the expression of the event is greatly limited, because such a diagram has no way to really express the spatial location of the object events and cannot show the object event happen and develop in space. These defects also greatly weaken the role of the Minkowski space-time diagram as tools for proper understanding of related concepts, and even cause unnecessary misunderstanding.

In order to solve this problem with space, the three-dimensional light cone was defined. This was later used with a large number of light cones, but although these diagrams appear to be three-dimensional, they're actually still two-dimensional and more likely to cause people to have misconceptions.

In Figure 4.2, the timeline occupies one-dimension. The rest is used to describe two-dimensional propagation of the greatest possible distance of object events. For example, we use light seconds to calculate the space in the figure. In the figure, go down one second from the "event (now)," emitted from -1 second of the time axis, at that time the event will only be propagated to the furthest 1 light-second with the radius of the circle at that second; at -2 seconds of the time axis, the event light propagation may reach a maximum radius of the 2 light-seconds circle, ... when you connect multiple largest possible circles spread over time, you get a three-dimensional cone of light as figure 4.2.

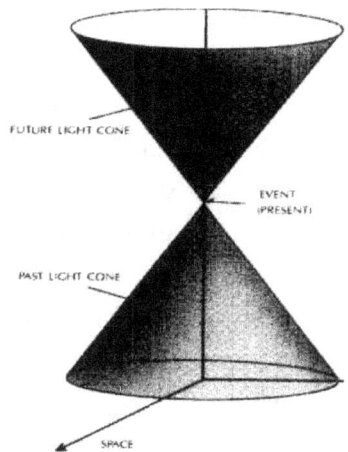

Figure 4.2 light cone

Then, by the same token, move upward in the figure from the "event (now)", you can draw the future light cone.

Please note that:

1. The actual light cone surface actually is only the possible spread distance limits of any event information. This is the light cone. So the light cone only defines the boundaries of event information dissemination, not the event itself, nor event information itself. Just like after the maximum speed of the vehicle is specified, the maximum distance possible the vehicle can reach can be defined at any time, but the information is not of the vehicle itself.

2. This space is just an abstract concept. So it cannot provide a quantified description. Instead, qualitative language to describe the image seems to become a direct representation. So it almost completely abandons the two-dimensional

Part I   A 4th grade math quiz that has been fooling human a century

space-time diagram of the "distance" concept. Just like after we know the maximum speed of the vehicle, we cannot describe any car-related motion information because there is no means to determine the expression of the specific location of the vehicle in the light cone.

3. Future light cone encircled "absolute future" is only an expected but not happened event at time "now." It correctly defines that if the "event (now)" has a "future," these future events will occur only inside this future light cone. But perhaps "event (now)" no longer continues to exist after the now moment. Then, how can the "future" come? Let alone "absolute future."

4. Description of the light cone object is vague. The "event (now)" P and the past and future events of the light cone are not clearly defined.

5. Hawking said, "If one knows what is happening at some particular time everywhere in the region of space that lies within the past light cone of P, one can predict what will happen at P." This is wrong. The future is basically unpredictable, even assuming that people have the ability to know everything related to the past (which is actually impossible). Even after mastering all the relevant information of stock movements, one still cannot know exactly in the next moment if the stock will go up or down. Do not exaggerate the role of the light cone.

Therefore, there is a need for a method to create an image that is both easy to understand and able to correctly describe real, four-dimensional space-time.

We've designed a method to depict objects and their events, such as multiple particles interweaving through space, in a four-dimension diagram. The three dimensional Euclidean space is used to unambiguously describe the locations of objects as they traverse throughout space, while a time axis, separate and independent from the Euclidean space, describes the time of moments when the objects of concern were at those points in space at that time. In contrast to the application example of the two-dimensional Minkowski space-time diagram, we can clearly see the four-dimensional space-time diagram is a simple, powerful expression of the development process of the object events that makes it very easy for people to understand what we are concerned with, the multi-event process of each object event occurrence and development, and the dynamic object development of inter-event relationships.

In order to more clearly express all the concepts, we clearly defined the event object, the event, and event information so that we can avoid the need to apply the light cone concept everywhere to express something different (event object, events, event procedures, light, and even the need to pass the information).

Here a simple image drawing problem. This is an example of figure expression of our four-dimensional space behind the event object world line. Let's try to see if we can use a light cone or similar conceivable way now to clearly express it.

Example: We want to set an arbitrary number on the Earth – assuming four fixed observation points, two positions closest to the Sun, and two at a position farthest from the Sun, for decades observing the Sun. Suppose the Earth's trajectory does not change. How can we express this? Can a light cone express it?

Let's take a few lines to draw the world line diagram with the expression of four-dimensional space-time and then summarize the formal definition of the four-dimensional space-time diagram. Finally, we will the four-dimensional space-time diagram to examine some specific application examples, including the one above.

FOUR-DIMENSIONAL SPACE-TIME DIAGRAM EXPRESSION OF THE WORLD LINE OF THE EVENT OBJECT

We use Figure 4.3 to represent the four-dimensional space-time of the event object world line, called the four-dimensional space-time chart of the event object world line, referred to the event object in the space-time world line diagram, and further can be referred to as the object space-time diagram.

The space-time diagram's mean spatial scale is above and the time scale below. The time scale corresponds to the spatial development process events.

Figure 4.3 is represented by the traditional space X, Y, Z coordinates.

Then draw a line from the past to the future, with the line parallel to the X-axis and beneath the traditional spatial coordinates. That is the time line. The current moment "now" is represented by the coordinate origin of this timeline, and the object past events draw in sequential order in accordance with the time occurred. We may on the above world line specify the time unit and scale.

The name of the event object world line is the name of the event object.

History of the event object world line is represented by the solid curves in Figure 4.3 of lines A, B and C. Sometimes, in order to meet the requirement of the application to be identified, a different name is given to the object event world line.

Event objects at time "now" is the cross intersection of the history solid and future dashed object event world line, and the location and moment the object event exists. The corresponding time is point "now" on the time axis.

Part I A 4th grade math quiz that has been fooling human a century

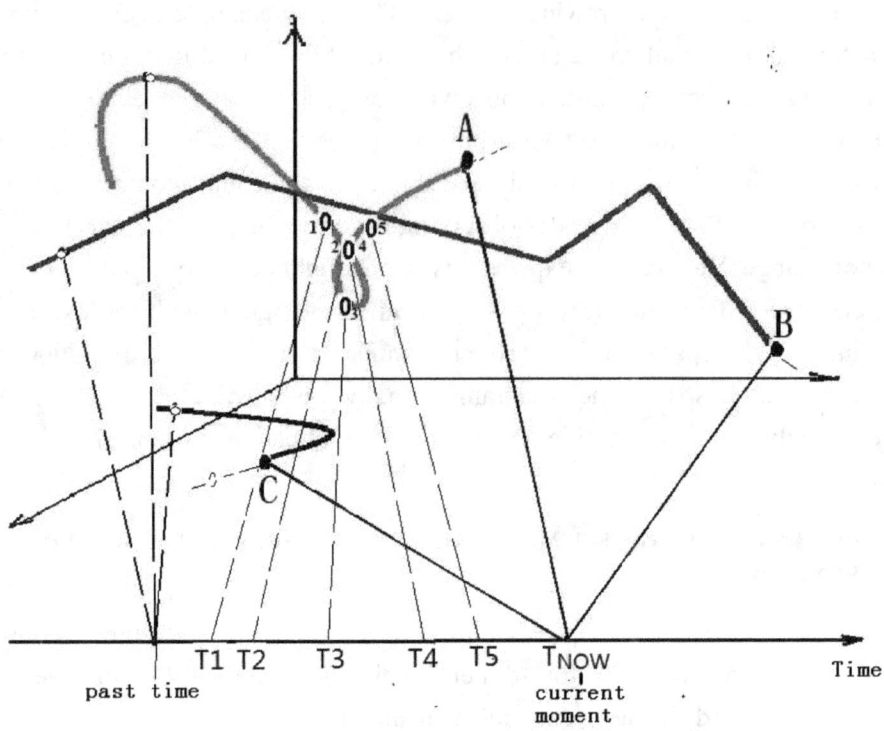

Figure 4.3 four-dimensional diagram of object event world line

Hollow dots on the world line represent the object events spatial position at some point in the past, but also the event object position ever existed, known historical events of the object, referred to the event. Points 1, 2, 3 and 4 are the points of C world line in the Figure 4.3.

The event object itself is represented by solid circles at the top of the solid history line. Time position "now" represents the spatial position of event objects at time "now," as shown in Figure 4.3 A, B, and C.

The possible future development path is represented by dotted line segments extending out from the object event solid circle. Typically, this imaginary dot line is very short because it is the expected development only, not a fact. The possible future event points we are interested in on the dotted line with our anticipated future position are represented by hollow circles (void center circles). This is shown in Figure 4.3 as the hollow circles on the dotted world line extending from C. You can also name the void circle.

Please pay special attention to the points on the world line C shown above. They correspond to the event-time paired sequence of events based on a combination of objects from the past to the present process occurring in strict order.

The time the event occurred of any point on the world line in the space other than the "now" is linked by a dotted line "moment Leader" that will connect the event points on the world line to the timeline. To make the screen simple, we can also not draw "time Leader," but use the names of the points on the world line and corresponding time axis to represent a "combined mark." For example, for world line C (1, T1), (2, T2) in Figure 4.3, the "moment Leader" and "combined mark" are drawn out. Usually, we use the same mark on the world line and the timeline for the same event.

"Now" is connected with a straight line between the object and event timeline. For example, on the C world line, the "now" moment is connected with the event object indicated in the figure (C, T_{NOW}).

In Figure 4.3 there are three lines A, B, and C, respectively corresponding to three world lines. The C world line is drawn in more detail, as we are concerned about the online historical events at points C1, C2, C3, C4, with each point connected to their corresponding times T1, T2, T3, T4, and the corresponding event object C at time "now" connected to T_{NOW}.

In Figure 4.3, there are two more interesting points that are the coincident points named 2 and 4 in the world line represented on the coordinates, but they do not coincide in their time marks of the incident. This requires two point coincident with the different order names expressed, and their "combinations marked" also has two groups.

In Figure 4.3, the event points on world line A we are interested in are 1, 2, 3, 4, and 5. The events of point 2 and point 4 are overlapping in space, but they occur at different times respectively, so their "combined marks" are respectively (2, T2) and (4, T4); There are two different corresponding "moment guidelines."

Thus, Figure 4.3 indicates the world line is able to represent correctly and clearly overlapping space-time object events even when the events, according to the historical time process of their development, correspond to different moments of this phenomenon. When we only use the "combined mark" without a corresponding "guidance timeline" drawn, the image will be more clear and clean, which is conducive to representing complex situations with multiple world line interactions.

The corresponding moments on the timeline in the above figure indicates that we can draw different timelines in accordance with different world lines, so we can clearly understand the events on the world line and express the complete event and its correspondence to a point in time we care about. Even the overlapped events in

Part I A 4th grade math quiz that has been fooling human a century

space can be accurately expressed.

For example, to study the correlation point we are interested in, such as the world line and points of events show in figure 4.4, the correspondence between the event and the time point of the event is represented as a basic event-time axis. The narrative corresponding related events can use the "combination mark" list and related text supplement.

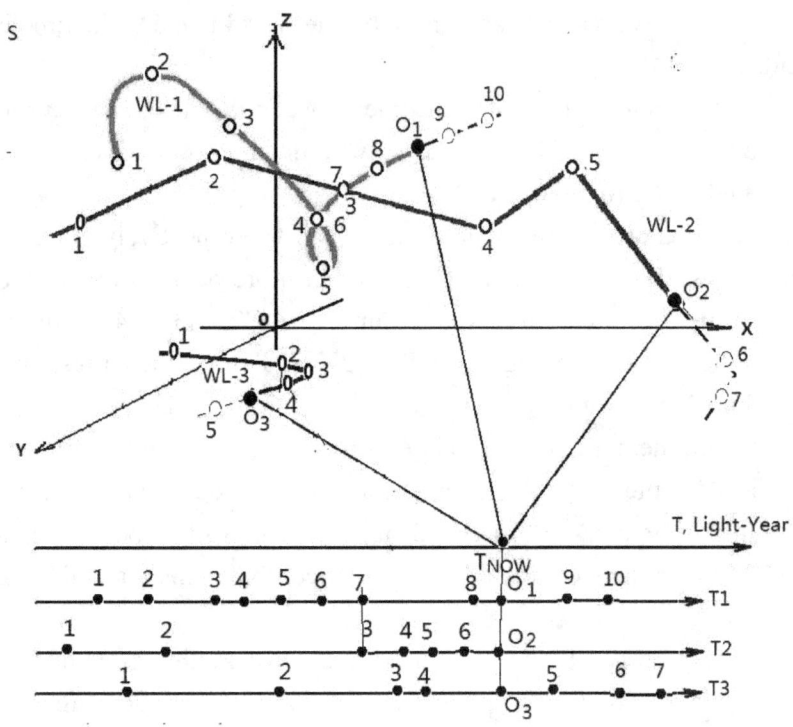

Events of object A, Point 4 and 6 are overlap, denote $A(A4, A6)$.
Event 7 of object A is overlap event C of object B, denote $AB(A7, Bc)$.

Figure 4.4 The mutual relationship between event objects on a four-dimensional object event world line.

Thus, we have basically completed the design sketch of object event four-dimensional world line representation. However, the story does not end.

Expand the four-dimensional space for the five-dimensional space, and even more dimensional space

First, we are expanding four-dimensional space to five-dimensional space.

If we have a need to express this, we can do it like in Figure 4.5: define an XY plane parallel to the X-axis, but with the space coordinates not interfering with the time plane. That time will be extended to the basic axis plane, and different events on the object and its corresponding time event space projected on the plane corresponding to the time of events on different world lines. Available in different angles and different events, the timeline is used to indicate the need to express the concept like Minkowski did.

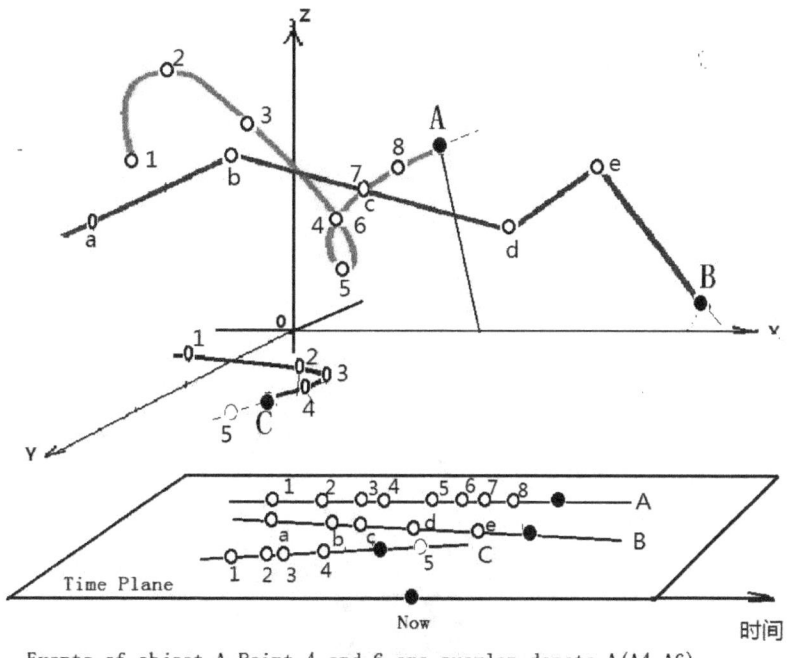

Events of object A, Point 4 and 6 are overlap, denote A(A4, A6).
Event 7 of object A is overlap event C of object B, denote AB(A7, Bc).

Figure 4.5 The mutual relationship between event objects on a five-dimensional object event world line

The possible methods to expand the five-dimensional space for the six-dimensional space: to event points on five-dimensional space, if we have more variables of interest, such as an event of interest related to the funding situation, then we can make a plan on the X-axis perpendicular to the plane parallel to the YZ plane, and make a Y axis parallel to the axis of funds E, this constitutes the XYZ-TT-E six-dimensional space-time-fund space.

The same approach can be used to expand the E-axis to EE-plane, and it would

Part I   A 4th grade math quiz that has been fooling human a century

become a seven-dimensional space.

On a plane perpendicular to the Z-axis, the same can be extended out of one or two-dimensional space. So we can get the final diagrams of eight-dimensional and nine-dimensional space.

In fact, if there is demand, time plane or fund plan can also expand into three-dimensional space. Not only can we expand the economic plane into three-dimensional views of space, the space can be used in the similar expanding timeline space... These may correspond to different levels of complex systems.

APPLICATION OF THE FOUR-DIMENSIONAL SPACE-TIME OBJECT EVENT WORLD LINE DIAGRAM

Now we use the space-time diagram to express an example of practical application.

In Hawking's "Brief History of Time" there is such an example as shown in above Figure 4.6.

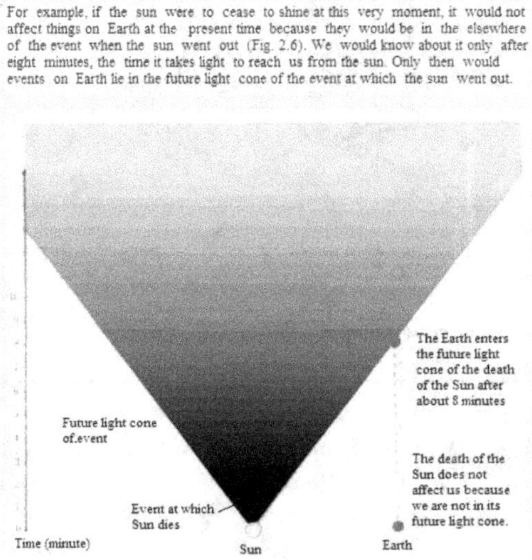

Figure 4.6 Hawking's example diagram of light cone application.

I do not know how to read this Figure 4.6. I do not understand it either. How did the Earth enter the dark event? How there is no point and event related before entering the relationship?

Before the Sun is actually destroyed, we know that the Sun has been shining on the Earth. After the Sun dies away, the sunlight continues to shine on the Earth, and the Earth does not right away know that the Sun has died. When the sunlight is completely extinguished, and the Earth receives no sunlight, then we know the Sun is dead. The causality is not shown at all in the light cone diagram in the figure.

Figure 4.7 is a four-dimensional space-time expression of the above mentioned sunlight going out event. This event involves two objects: the Sun and the Earth. The Sun at the "now" time represents the cessation of sunlight emitted by the Sun. The Earth rotates around the Sun while the Sun's transmissions are slowly changing, until the Sun finally disappears from the Earth.

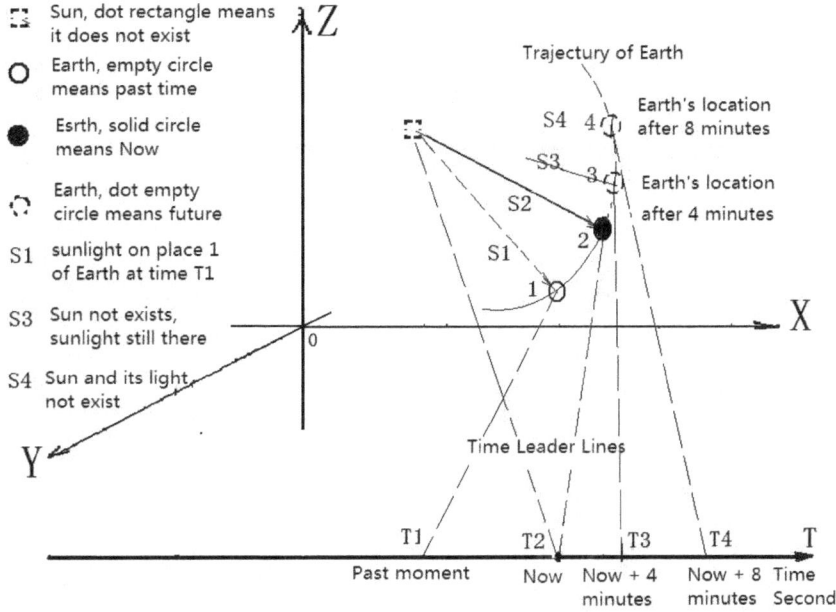

Figure 4.7 Hawking's example diagram of light cone application.

If we have more content to draw into the map, or to make the image clearer, we can remove the "moment Leader" lines, instead using "combination labels" to indicate the same information, thus obtaining the following Figure 4.8.1 and Figure 4.8.2.

Part I   A 4th grade math quiz that has been fooling human a century

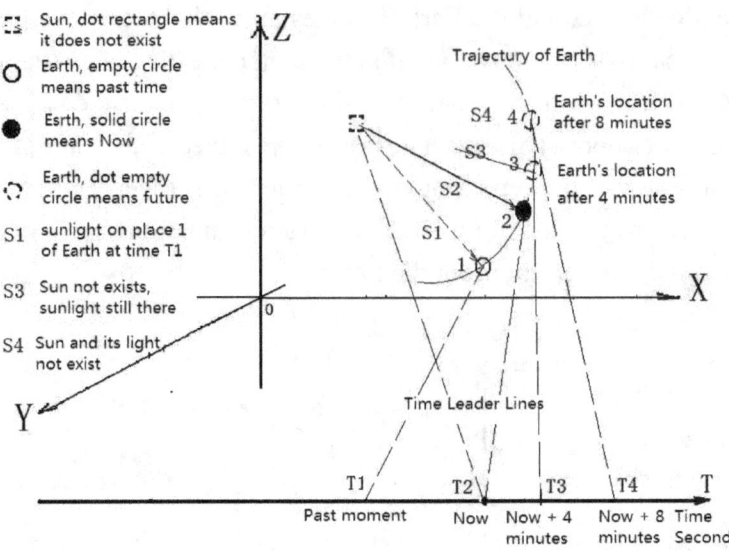

Figure 4.8.1 Describes the event "Eight minutes after the Sun is dead, the Earth will know that event" with a four-dimensional object event diagram

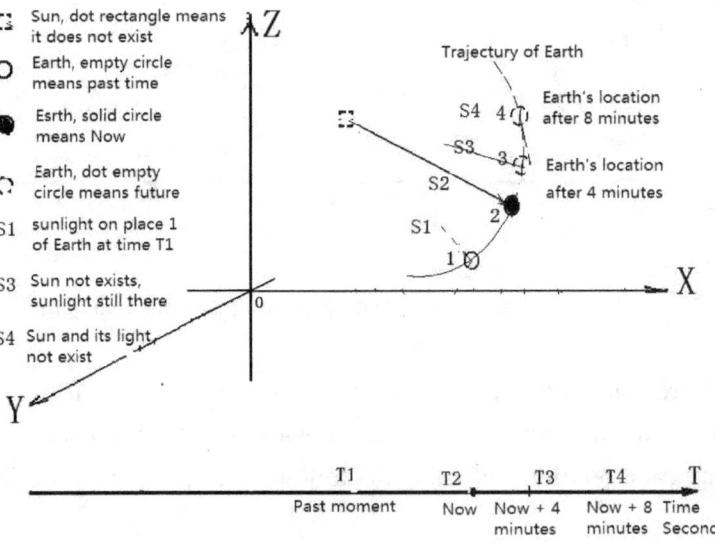

Figure 4.8.2 A space-time object event world line to depict "Eight minutes after the Sun is dead, the Earth will know that event" with no "time Leader" lines in the graph. The Earth orbit points we are interested in "combined mark" notation are (1, T1), (2, T2 = TNOW), (3, T3 = TNOW + 4 light minutes), (4, T4 = TNOW + 8 light minutes).

Now we summarize the definition of the four-dimensional space-time image of the object event world line in Figure 4.9.

Four-dimensional Space-Time Diagram of Object Space-Time World Line

Space: Usually using orthogonal X, Y, Z coordinate system of the conventional Euclid space

Time Plane: parallel to the XY plane and space, perpendicular to the Z coordinate space, and space systems disjoint Independent flat

Basic Timeline: the traditional one-dimensional time ruler with arrow, and even the definition of the unit of time scale and the "Now" moment as origin point.

Object: the presence of a substance in the universe, the celestial, people, matter, particle, etc. we are interested in our study.

Object World Line, a line of an object in 3-dimensional space with each point corresponding to an event of the object. This point also related to one point at Time-Axis. The points are arranged strictly sequentially to the time from the past to the future, with the original point on the Time-Axis related to a event of the "Current" moment of the line, past events are happened before the "Current" moment and future events may perhaps happen or not happen after the "Current" moment. Past point is denoted by solid circle and future points are denoted by dotted circle.

Time Line of the world line, one-dimensional way axis, is the projection of world line on the time plane in time sequential. A world line corresponds to a time line. If the image is not confused, the basic time axis can also be use as time line of the world line.

Events on the world line:: the point we are interested in the world online, also known as event points, or event.

World line event collection: a collection of all points of events on the world line studied.

World line collection: all of researching world lines collection.

Time Line set of world line: all of researching time lines of world lines collection

Time line plane (five-dimensional space-time diagram): define a time plan parallel to XY plane but without any intersection with the world line space. Extend the time line to the time plane, project events on different object world lines on the time plane, with different angles in different time axis to represent the concepts to be expressed.

Six to nine-dimensional space: on a plane vertical X axis and parallel to the YZ plane, draw an E axis parallel to Y axis, thus construct an XYZ-TE-E 6-dimensional space-time diagram. Following the same method, the E-axis can be extended to EE plane, thus gets XYZ-TT-EE 7-dimensional space-time diagram.

Figure 4.9 expression for the four-dimensional object event space-time world line

Now back to the previous example. It should be easily solved, right?

Example: We want to set an arbitrary number on the Earth – assuming four fixed observation points, with two positions when the Earth is closest to the Sun, two when the Earth is at a position farthest from the Sun, and observing the Sun for decades. Assume the Earth's trajectory does not change.

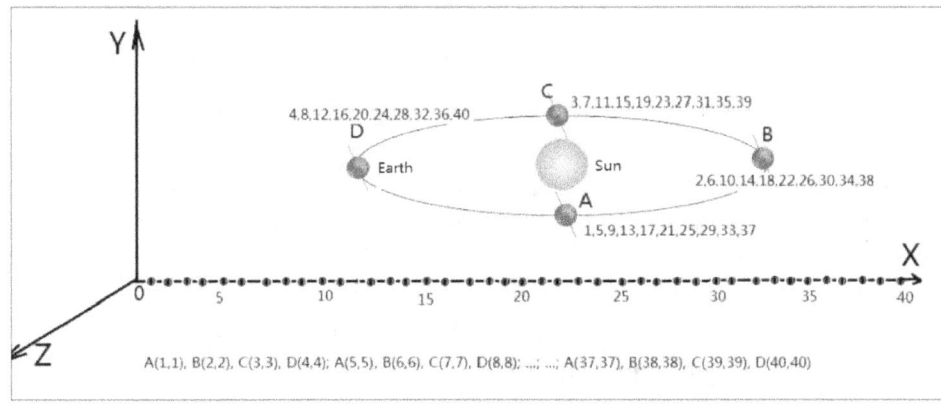

Figure 4.10 Four-dimensional space-time world line for observing the Sun example

Draw an elliptical Earth orbit around the Sun, and the Sun is a fixed point to the Earth's orbit. Find the four points, marking the order 1,2,3,4, repeat marked 5,6,7,8 ..., 39, 40, until after a decade is marked. Then turn to the time axis and label 1, 2 ... 39, 40. If we need to indicate the specific events, we can record these events

Part I A 4th grade math quiz that has been fooling human a century

under the same label, etc. For example, A (1, 1), when the need to record relevant observations can be recorded as A (1, 1, observations -1)....

A third example is shown below in Figure 4.11. Two high-speed aircraft A and B are flying at altitudes of 10,000 and 5,000 meters in the X direction respectively. In the four-dimensional space-time diagram world lines their movement is marked as events. These two mutually independent world lines will not be arbitrarily change because of external factors, and the world line of B would not be changed by the reason that we observe the relative movements inside B from A.

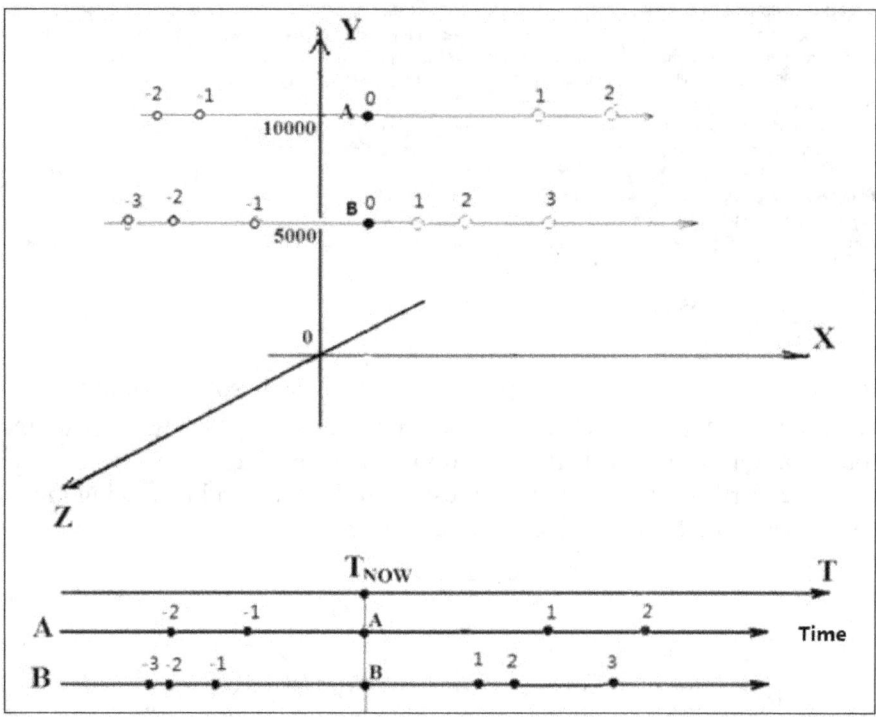

Figure 4.11 Two high-speed movements of the aircraft A and B, respectively, in the X direction of flight at 10,000 meter and 5,000 meter altitudes. Similarly, we can draw Einstein's stationary system and dynamic system in the same way, and obviously the relativity problems of any non-simultaneous nature do not occur

The Explanation of the "Object Existence Law" using a Four-dimensional Object World Line Space-time Diagram

The existence theorem is modified from the chapter of "Time Theorem" in "Who Should Talk about the Cosmos", with a lot of changes. For one, the light-time law has been changed to the object existence law. Because it was the integration of all my own ideas, and the existence of the law is one of the most important ideas of the author, these three sections are modified and integrated here.

Now, on to what in the original book was called "Space-time Is Not a Part of the Existing Universe" with a new awareness. Interested readers can compare one with the other.

Object Existence Law: Objects Exist Only at Current Moment "Now"

Definition of the Law of Object Existence: objects exist only at the "now" moment. The **past of an object** is the memories of sequential incidents the process of the object's evolution left to us; the **relative future of an object** is the event of the object that has already happened, but its information is transmitting to the destination and has not yet been received, and the process of this event is ordered and is irreversible; the **absolute future of an object** does not exist at this "now" moment but only as prediction or expectation for the event world line evolution trajectory of the object events at specific future moments. Watch something external to the system from any place to see the object itself exists only in the corresponding system itself at the "now" moment. All the rest occurred in the history of the event records before this "now" moment, and events after this "now" moment do not exist, cannot be seen, and possibly would happen in future time.

This is the **Object Existence Law**, referred to as the **Existence Law**.

Here I define relative future and absolute future new concepts. For example, working in the city, Peter wrote a letter to his wife at home. When this letter was put into the mailbox, the "Peter's wife is getting a letter from Peter" became a relative future event to his wife, because he has written the letter but the letter has not yet made it to the hands of the recipient. If Peter has not written the letter, then the " Peter's wife is getting a letter from Peter " is the absolute future because it has not happened yet. But we know that based on the historical record, there is a big possibility that it will happen under normal circumstances.

The relative future is what has happened, but the information about the happened event has not been transmitted to the recipient. If there is no problem with the transfer process, the event finally occurs; but also possibly that may not occur. If Peter wrote a letter to his wife, the event of "Peter's wife is getting a letter from Peter" became relative future. But it were also possible that Peter's letter would be lost in the delivery process, the event of "his wife is receiving the letter" will not occur.

Part I   A 4th grade math quiz that has been fooling human a century

The absolute future belongs to the event did not happen. Peter has not written the letter yet. In the future, it is also likely to occur or may not occur.

The concepts of relative and absolute future are important when discussing the issues of the universe. The sunlight emitted five minutes ago, to the observer on Earth, it is relative future event because it will be received and be able to be observed on the Earth for more three minutes. The sunlight emitted by the Sun 10 minutes later is the absolute future even to the observer on Earth. We believe that the Sun will continue to exist after 8 minutes and continue to send Earth light; there is a large possibility that "Earth people received the sunlight that emitted by the Sun 10 minutes later" will occur, but it has not yet become a reality. The Sun may have been destroyed in the ninth minute.

In my book "Who Should Talk about Cosmos", I proved the Existence Law with narrative language. Here, again, I prove the law using the object event world line and then make a more complete description.

Object existence Law in the world Line Description

A boat glides slowly in a calm lake. The boat is the point the object now exists; the wave behind the ship is the path the ship ran, and is where the boat once existed in history; where the front of the boat will reach in future, we can probably calculate out and expect when to arrive at according to the current position and history of trajectory, but that has not happened yet. Perhaps in the next moment, the boat will sink or change its direction.

The world line concepts we use here has a clear definition: an independent object represented in a world line, with the beginning of the world line corresponding to the event start position and time. The moment "now" corresponds to the world line forward vertex and, also, is the existence location of the object. All event points on the world line are strict, not chaotically arranged, from the beginning to the end point between the vertices, from the past to the present moment in orderly sequence. Each point in turn is the expression of the state of the object event at that time, and in turn, from the moment they occur to now corresponds to each point on the timeline.

From the viewpoint of collections of objects, a planet can have its entire trajectory. All the time, it only exists at the forward end of the trajectory; its past is the memory of the locations that it has moved through, and its future, based on expectations made from its past and present, does not exist in reality and is not shown in the figure. Of course, we can also draw "future" by the dashed line in the back vertex, behind the corresponding point time "now" (more right) on the timeline.

A planet can be broken down into the mountains, rivers, humans, animals, birds, insects, vegetation, and sediment... For every one of these, we can depict the individual based on knowledge of the individual's world line. The forefront of the world line is the existence of the single individual at the current time "now." The collection of all of these separated individual object's world lines, is the planet's existence at time "now."

We can gradually, finely decompose, and can also gradually polymerize, from large to small or small to large, but there is only one set of world lines on their apexes represents the existence of the real objects!

The Chapter Conclusion

The Minkowski diagram at the beginning of this chapter was made many years ago to express Einstein's relativity concept. For centuries, people have been using it to express various concepts of relativity. But because it is only a two-dimensional image, which only represents one-dimensional space, it is neither intuitive nor easy to understand.

More serious is that it introduced a paradox and a disaster into the relativity theory. We have discussed the concept in detail.

Using four-dimensional, five-dimensional, or higher dimensional image representation, we overcome the shortcomings of Minkowski's two-dimensional image, and make it possible to better understand more clearly space-time and related concepts.

Space, time, is forever a mysterious theme.

Part I Conclusion

In this part, we have three important experiment designs for testing the correctness of the relativity theory. They are:

1. The cardboard bends the lamp light experiment, and the observation design for the Moon bending the sunlight experiment. The experiments try to quantitatively determine if the bend light is caused by the gravitational field or caused by the diffraction effect.

Part I   A 4th grade math quiz that has been fooling human a century

2. The experiment of clocks combination to check the "moving clock slower" paradox

3. Verify the "moving meter stick shorter" disaster combination sticks experiment.

If we can finish these experiments seriously, then the knowledge of relativistic theory will be on a new level. The first experiment has a better chance of success.

The theory of relativity, which changed the thinking of human beings and ruled the scientific community for nearly a hundred years, turned out to be an erroneous concept that had not been rigorously proved by theory and experimentally, and was based on the primary school arithmetic of speed superimposed model.

The theory of relativity was first established on the basis of "only few people in the world can understand" statement from several famous scientists. And the observation result of the star light deflected by the Sun during solar eclipse affirmed the relativistic theory, established its unshakable unquestionable position of highest science god. Is this not the most peculiar scenery on the road of scientific and technology history? Should not historians of science and technology take a good look into the causes of it?

Science will progress.
Cosmology will progress.

There are so many obvious issues that need to be thought about and resolved. For the scientific truth, we should carefully think about them.

I did a little bit of thinking, squinted a little to open the eyes to the desire of the truth, so there is a "mathematics is old" lament at the end of the book.

It is not that genius is not being generated from the world, but in the industrialized world, there is no land with soil for genius to grow in.

Being born a genius makes it difficult to grow up, and it's even more difficult to grow up with new ideas.

Ills of the times!

PART II
THE BIG BANG THEORY CRITICISM

The imagination of the Big Bang theory can be called both fantasy and science fiction.

- Why the night sky is not bright? Crack the Olbers' Paradox for hundreds of years without assumptions.
  One light-year mystery - if the Sun is moved one light-year away from Earth, on the planet to get the same intensity of the sunlight, we need to add about four billion suns; "two suns" case; boundary of dark space and bright space; how quantitative describe the "dark" of night? Calculation dark space and light space.
- Looking for planets humans can move to; search for alien's base; looking for the hidden celestial objects, and dark matter.
  One light-year mystery - Flying over one light-year mankind needs thirty-seven thousand five hundred years; define hidden celestial objects; all the people who thought it was something dark matter, and the vast majority of dark matter that people are looking for are hidden objects. The definition of the mysterious dark matter is against the basic principles of science.
- The speed of celestial image: Subversion celestial red shift theory.
  We need pay attention to this: There should not be a huge difference in Doppler redshift values between different instruments of different sensitivity; Over time, the Hubble Constant has seen a huge difference in values. In the historical debate held on the big differences of numerical Hubble Constant values, the two factions fought over whether the Hubble Constant is 80 or 40, double the difference. This makes people wonder whether the redshift is actually caused by the Doppler Effect. So many scientists, for such a long time, were confident that their observations were

correct. Then who was wrong? Can we switch our thinking method in such a way: Everyone was right! The theory behind the interpretation of these observations was wrong! Correct theory should be able to explain all of these different observation results obtained from different telescopes! And in deed our theory of observed image speed changes can explain everything about celestial red shift!

- Observation test design for verifying Redshift theory.
- NASA whole-sky microwave background map questioned: we cannot apply data measured on a sesame to the whole Earth.

  No stationary electromagnetic waves; electromagnetic wave propagation movement must comply with the basic physical laws; don't consume Copernicus's law at will.
- A century bet with NASA: 100 years cannot find Dark Matter that more than hidden celestial objects (HCOs).
- Sometimes seeing is not true. Distant universe photos obtained by Space Telescope almost all are false!

  Aspect of a million light-years distant star photo, each point in the figure represents different ages of 10 thousands light-years.
- Reject the Big Bang smoke fog.

CHAPTER 5

TANGLE BETWEEN BRIGHT AND DARK
– Why Isn't the Night Sky Bright?[3]

Father and son are looking at the clear night sky. In the dark sky, there are countless stars flashing.

"Dad, do you know why the night sky is dark?"

"Because the Sun went down."

"But there are so many stars, why can't they light up the night sky?"

"Because the stars are too far away from us."

"Oh, I see! They are too far away!"

Amazing Thinking Turned into a Big Problem – A Brief Introduction of Olbers' Paradox

Sitting on the steps of Heaven, the night darkness wraps me like cool water while I'm watching the stars Altair and the Vega that are a couple separated by the star river.

Beautiful night sky, mysterious night sky, awesome night sky, physically and mentally desirable night sky...

Throughout the ages, facing the sky, numerous heroes have sung of its vastness and sighed at its deep mystery; countless writers and poets have sought to express the night sky, brewing their wonderful thoughts and beautiful poems...

---

[3] Refer to appendix 1. "New Solution on Olbers' Paradox: Why Sky is Dark at Night?", Matter Regularity 2015 2, Vol.15 sum No. 68, p9 - p19

The night sky is like a black silk inlaid with sparkling diamonds. If I could gently flick the silk, the stars in sky everywhere would rattle its song...

The German amateur astronomer Olbers gazed up at the night sky with numerous thoughts:

Why is the night sky black?

How can the darkness possibly beat the light?

The night sky should always be bright!

The amount of light emitted by the numerous stars, when combined in the night sky, should be bright enough to overcome the darkness. But the reality is that the dark rules the night sky. The bright should overcome darkness all the time, but the reality is: dark at night, victory over the light!

This is Olbers' paradox.

The simple explanation of Olbers' paradox is as follows: if the universe is infinite, is in steady state, and stars are uniformly distributed, then the night sky should be bright rather than dark.

The full name of Olbers is Heinrich Wilhelm Matthias Olbers, an authentic Deutsch name.

Olbers studied medicine in Gottingen, graduating when he was 22 years old. After graduating, he began practicing medicine in his hometown of Bremen. He earned his living treating patients during the day, but his real passion was watching the stars at night, and he threw both his money and himself into it. (For this foolish behavior, I would have liked to cast several criticisms, but then I discovered that I was the same kind of fool as Olbers. For almost ten years, I have struggled to write and calculate about the void of space. I have devoted such a big chunk of my spare time, including during the fireworks of New Year's and under the lights of Christmas, that I realize that it would be difficult for me to enjoy any other kind of life.)

Olbers converted the top of his house into an observatory. Though still young, he did not love beauty but loved stars. The star shining in the sky became his field of joy every night.

He made tireless efforts, year after year, in exchange for fruitful returns. He was the first to design a satisfying method for calculating the orbits of comets. At the age of 49 (1807), he discovered an asteroid, giving it the name "Pallas." Five years later, he found another, similar asteroid, and he asked his friend Gauss to name it: "Vesta." Olbers was the first to propose that these objects came from an "asteroid belt," and this became the official name used by the astronomical community. At the age of 57, he discovered a periodic comet, and it was named after him (Official name 13P/Olbers).

In 1823, the 65-year-old Olbers still opened wide his anxious and childlike eyes to the now familiar sky, thinking abstractly and freely about the problem of the dark night sky. He thought that the sky was like a big ball filled with stars, galaxies, and various light-emitting substances, both visible and invisible. (In this book, I use the term "stars" to refer to all of these objects collectively.)

In reality, we have a dark sky at night. According to theoretical proof, however, the sky should always be bright. This conflict of the practical against the theoretical constitutes a paradox.

Olbers theoretically proved that there should be no dark sky in the following way:

Suppose the universe is infinite and the stars are evenly distributed in the space.

In his imagination, Olbers divided the universal sphere into imaginary circles and layers so that it had an onion-like structure; i.e., from the inside to the outside, he divided the universe into equal intervals of countless layers.

Now take any layer, for example, the 10th layer, for the proof.

The distance of the tenth layer from our telescope is ten times of the distance between the first layer and the telescope. The light emitted from each star at the tenth layer will be one hundred times weaker than the light from the same star at the first layer. This situation, actually, was explained in the preface section of this book, with a formula explanation.

However, from the perspective of the number of stars to calculate, the space of the tenth layer is a hundred times larger than the first layer because the stars are evenly distributed in a spherical space, the number of stars at the 10th layer is a hundred times the number of stars at the first layer!

So, at the tenth layer, although the brightness of each star attenuates with the square of the distance so that it is 100 times dimmer than at the first layer, the number of light-emitting stars at this layer is 100 times greater than the number at the first layer.

Light emitted by each star is weakened at the tenth layer, but the number of stars is increased at this layer, and both are 100 times. The luminance of a single star attenuates at the same rate as the number of stars increases. The increase or decrease of the brightness is offset by the sum, and the final result is the same: the amount of light that reaches our eyes from distant stars at the tenth layer is the same as the light from those at the nearby first layer.

Remember, this calculation of the tenth layer can be applied to any other layer of our onion-like modeled universe. The result is the same: the light comes toward our eyes from any layer in the same amount.

Thus, assuming that we can put ourselves at any observation point in space, no matter where in the universe, every layer, both far or near, ultimately contributes the same level of brightness to our eyes.

The result of this calculation is:

No matter where we are in the universe, no matter how near or far away the stars are from us, and no matter what time it is, starlight should be as bright as daylight, provided that the universe is infinite and the stars are uniformly distributed.

The above mathematical model Olbers used in his proof is called the **Shell Layer Model.**

The Shell Layer model completely rejected the idea of using "distance" to solve Olbers' paradox.

The Shell Layer model used mathematical proof to show that, if the universe is infinite, eternal, and static (actually, it should be said that the universe cannot be expanding) and the stars in the universe are uniformly distributed, then the night sky should be as bright as day!

But why, then, is the night sky dark?

In 1823, Olbers wrote down his thoughts on the subject in a scientific paper and published it.

This paradox embarrassed him, and also embarrassed all scientists for nearly two hundred years – this calculated result that light should illuminate every corner of the universe was beaten by the reality of seeing a dark night sky – and it was called Olbers' paradox, also known as the mystery of the dark universe.

Olbers paradox states: theoretically, the sky should be forever bright, but the reality is that the sky is dark at night.

Cracking Olbers' paradox requires theoretical proof for the existence of the dark night.

Seemingly simple and naive, it might seem now that the one phrase "too far" should have been able to veto the issue, but the paradox turned out to be a big problem over the centuries!

For years, hundreds of astronomers directly or indirectly attempted to solve this problem. But each solution was either subsequently rejected or was hardly convincing in the first place. And almost every historical solution was based on certain assumptions, such as that the universe is full of some kind of substance, or that the universe is expanding, and the like.

What we will provide is a completely non-hypothetical solution!

One Scientist after Anther Attempts to Solve Olbers' Paradox

In 1600, under the Ming Dynasty, the Chinese nation was being ravaged repeatedly by feudal decadence, stupid and greedy bureaucracy, desperate peasant revolts, and barbaric nomads. The country was torn apart, going from the world's richest nation, with the highest GDP, down to a humiliating period as a conquered people.

In the West, however, in 1576, western scientists like the British astronomer Thomas Digers were thinking about scientific problems such as why the sky was not bright.

Olbers first introduced the previously discussed Shell Layer model as a theoretical denial of the use of the "distance concept" to answer the paradox. All subsequent solutions followed on that history; basically, no one again tried using the "distance concept" to solve the paradox.

I have to say that this is one of the negative impacts on history brought on by Olbers. He raised the issue, but it's a darn good reason to block the right way to solve the problem.

I have often thought that many cosmological problems could be solved if the "distance concept" could be used, but no one seems to follow up on the idea. Universe theory does not seem to pay attention to the "distance." It seems that no matter how far away, no matter how weak the light from more distant objects is, the information can be received. This perhaps started hundreds of years ago with the negative impact of Olbers' proof against the "distance concept."

But from the Limit theory presented in the preface, we already know that the light from far away will eventually lose all celestial glory!

For centuries, many famous and lesser-known scientists have tried to answer this paradox from all angles but haven't been able to find a satisfactory solution. From their answers, we can see the impact of Olbers' ideas; no matter what solution is tried, it always either includes the entire universe, considers only one or two data points to represent the brightness of the whole universe, or focuses on computing background brightness for extragalactic space. But very few scientists have ever pulled back their focus from the distant universe to look back at the Earth, the Sun, and objects near them. Most seem to think that, after determining a certain brightness or temperature in some area of space, they can solve the problem over the whole universe, for example, determining the background brightness of the galaxies of space and then using that to solve the problem of brightness for the whole universe. From observations of near-Earth space, and from just a few years of data, they try to draw whole sky map for the entire universe.

However, the distribution of stars in space near the observer, in fact, is the most important prerequisite in determining if the skies are dark or not.

For example, suppose that the space near the Earth has two suns and that the Earth is in the middle of them. Well, would there ever be darkness on Earth? Would people on Earth still see the night sky? And if these two suns were on the same side of the Earth, then would night still come at the end of every day?

We call this the "**two suns problem**."

In reading the following various historical solutions, please compare them with the "two suns problem." The comparisons that you find will allow you to draw some interesting conclusions.

AN OVERVIEW OF OLBERS' PARADOX

Olbers was not the first person to eat this crab, but he was the first one to put the paradox into formal thesis and prove his argument with the Shell Layer model. Prior to him, a similar question had been raised directly or indirectly by many scientists. The question of why the night sky is dark is a tempting one, and since ancient times, has posed an interesting problem that wise philosophers could not help thinking about. Of course, if a person has never heard of this problem, then he or she has to be very smart to imagine it in the first place. He or she needs to be wise, have enough knowledge, and have some kind of romantic feeling.

It's like evaluating a graduate student for quality and talent. One of the main criteria is to look at the level of the points he raised. His points should not only present new ideas, they should also be based on sound reasoning and sufficient proof. The fundamental difference between whimsical and cranky is that the former is the mark of a shining of genius, while the latter is characteristic of an idiot.

Olbers' Paradox is a problem for hundreds of years, before Olbers formally published in the form of academic papers, the issue was indirectly raised ; thereafter, many scholars have proposed a variety of solutions, showing a flourishing trend.

But in recent years, in the Big Bang smoke filled, only two voices: the presence of the night is because of the expansion of the universe? Or insufficient time of the cosmic wave propagation prevents the bright dominating the space? The following sections will introduce the history solutions in detail.

In fact, the two theories are essentially the same: using the Big Bang from different angles to prove Olbers' paradox then turned back to become the evidence of the Big Bang. This "scientific" mutual self-certified argumentation is not worth promoting.

THE BIG BANG MODEL AS PREMISE OF OLBERS' PARADOX SOLUTION

Expansion of the universe of the Big Bang theory may well explain Olbers' Paradox.

In the expansion of the universe of the Big Bang model, the universe is not infinite, there is no superposition of numerous layers faint starlight of the shell model. And because the universe is expanding, light from the viewer from a distance of stars issued, cannot reach the observer; and the light from vast distances reaching the observer will be diluted in the propagation process; far more than a certain distance Starlight will not reach the observer. Thus, the lights reach the observer are limited, and are not so strong, the night sky will not be covered by bright.

Look, how perfect explanation!

So the Big Bang theory is by far the most reasonable explanation can explain Olbers' Paradox - the night is there - one of the two most popular argument of contemporary cosmologists generally accepted. While conversely Olbers' Paradox is one of the powerful evidence of the Big Bang theory.

Since the Big Bang model is based on expansion of the universe, therefore, to clarify the distinction, highlighting the relationship between Olbers' Paradox and the Big Bang, the layer of the shell model proposed by Olbers needs to be clearly marked, layer shell model of the universe is a "static state" (in this way, if the expansion of the universe is denied by the paradox, it certainly means that the Big Bang theory is correct).

Therefore, there are three conditions of the shell layer model: that the universe is infinite, is in "static state," and astral evenly distributed in the universe.

Now we see about the document on Olbers' Paradox narrative (such as Baidu science and technology in Olbers' Paradox entries) generally is written like this: if the universe is infinite and is in static state, the night sky should be bright. In fact, this description is not complete, also need to add a uniform distribution of stars in the universe, because it ensures that in the shell layer model the number of stars in each layer is increased proportionally.

So Olbers' Paradox more accurate description would be: if the universe is infinite, is "not expanded", and uniform distribution of stars in the universe, then the night sky should be bright.

Change the "static state" with "not expanded" is very important!

Because the universe cannot be in "static state". All stars are in accordance with their own running track. "Static steady state" in the name is easy to make people understand astray, even pupil know there is no 'static steady state" in the universe, all things are in motion, say the universe is "static steady state" theory is absolutely false theory. We say that the universe does not expand, does not imply any that the universe is in "static steady state": We know that the planets turn around the star, the solar system rotates around the center of the galaxy at to 250 kilometers per second speed, and the Milky Way also has self-rotation. Even if the universe is expanding, these movements are still engaged. But if there is no expansion of the universe, there is no Big Bang.

Based on the expansion of the universe, the Big Bang model as a solution to Olbers' Paradox is persuasive, so many other non-Big Bang solution, have largely been blown up.

However, think and compared to the "two suns" problem, the Big Bang and a limited time solutions both are not perfect!

Our solution without any assumption will be given is: whether the universe is finite or infinite large, and is expanding or not expanding, there is always a bright sky and dark sky alternately, thus Olbers' paradox is not established. Thus, Olbers' Paradox will not be one of the Big Bang supporting theory.

Before explain our no assumption solution, let's look at the history of those schemes that were blow off by Big Bang and the two still surviving solutions recognized by present cosmologists.

Do not forget ponder compare "two suns problem" while reviewing the history of each program.

Brief Review of part of Historical Solutions

Here first we introduce two people used intuitive methods of distance before Olbers. The longer the distance, the lower the brightness, which is the most intuitive, before widespread use of Olbers. But since Olbers in the paper of the shell layer model gave a formal denial of the "distance" in this paradox solution possible role, people will have no argument with the distance since then.

However, I will give a proof back in surprise. Later in this chapter, I will use impeccable computing , break the hard shell of the distance layer shell model by Olbers, so that the "distance" role in solving the problem of the universe (not just to solve Olbers' Paradox) on the big stage once again become the protagonist of a grand debut!

In 1576, British astronomer Copernicus inspired by Thomas Digers new geocentric theory, the first time described the universe without borders, in this universe of stars is no longer tied days dome on (the ancient Greek scholar Aristotle is so described). Infinite light endlessly toward the sky. As they become more sophisticated, their numbers seem less and less, until we can no longer trace by eyes. Huge distance makes the most of these rays are disappeared in front of us.

In the narrative is not just appear endless stars, also it looks hazy expressed planet is not endless thoughts. Diggers on distance, as well as his later ones, including the famous scientist William Gilbert (1600) and Gallieo Gallilei (1624), have in vain with the distance argument fuss. And these are denied by Olbers layer shell model.

In the decades before Olbers, Kepler although not explicitly raise this paradox, but have explored this issue in 1610 of distance-based solutions. But his distance is "limited distance" that space stars look like as an island group that is a limited universe. Kepler finite universe, in fact, changed the Olbers' Paradox, opening the dispute to another question: the universe is finite or infinite? The Big Bang theory is clever than Kepler's answer because that answered why the universe is finite. Although the fact is the pot calling the kettle black, what is there before Big Bang and out of the island is always no answer.

Due to use the distance to solve Olbers' Paradox was failed, people turned to other possible factors.

Non-uniform sources: In 1721, Edmond Halley wrote that if the number of fixed stars is infinite, then the whole sky would be the bearer bright. These statements clearly say although Olbers' paradox named after him, but Halley was more than he would have made a similar initial concept.

Indirectly, Newton when trying his new theory of gravity, also need to have an infinite number of stars evenly distributed in the boundless universe. If space is not perfectly uniform distribution, each star in some way would be uneven gravity, thus the entire Newton universe will collapse.

Absorption: The most famous theory to solve Olbers' Paradox is that he proposed the theory of propagation of light in the way to be absorbed. Olbers hypothesis space between the stars of the universe is not transparent, but a thin layer of material. This substance will absorb the light emitted by the stars, the stars farther distance the more light is absorbed. Olbers detailed calculation on light emitted by the star Sirius Case and concluded that stars are superimposed light cannot light up the dark night sky. Olbers use it to explain his paradox. But this theory in 1848 was rejected, because there is less space even layer of material, thus the propagation of light is absorbed on the road, but the absorbing light substances will be heated and emitting light again. Finally, the dark night sky is still illuminated.

Fractal Universe: Herschel in 1848 made a completely different explanation from the traditional idea: the light source is divided into mufti-level or a fractal, which he called "principle of subordinate grouping." so that stars are grouped into galaxies, galaxies into clusters, clusters into super clusters, and so on, literally without end. In this model of the universe, with the increase in volume, the average density continued to decrease, its limit tends to zero. As a result, such an infinite universe cannot touch a star in an infinite number of directions. But the success of this model requires a fatal premise: the size of galaxies tends to be theoretically infinite. If, for example, the size of the galaxy has a limit, then the "cluster" would merely replace the Olbers' layer shell model in the position of the stars, and the fractal model of the universe would return to become layer shell model.

Tired Light: In 1707 N. Hartsoeker suggested that the sky is dark at night because the intensity of light drops off more quickly than expected over cosmological distances. During 1920 and 1930, represented by F. Zwicky, do not want to interpret astronomical observations redshift for expansion of the universe, made the theory of tired light. This theory was disproved later by observations. Also the tired light theory have encountered the same problem immediately below to introduce, the absorption theory Olbers used. Using observations to deny a theory in astronomy is very common, but it is a wrong way of thinking. Because the observed results are basically wrong. With the wrong observations to deny a theory has been common mistake of astronomy community. We will prove this in more detail later.

Dark Star Theory: F Arago (1857) and E, E, Fournier d'Albe (1907) said that the black night sky due to the distant starlight is blocked by solid celestial objects. This theory was soon overturned by discovering yellow, red dwarfs, black holes and dark matter in modern cosmology.

Overduin & Wesson (2009) considered the finite time theory is basically right. This theory contains two aspects: finite speed of light and limited light propagation time.

At that time, Einstein has not raised the speed of light theory. Surprisingly this finite time theory was raised by American poet Edgar Allan Poe in his poem. He wrote in 1948, "a prose poem" in: "Were the succession of stars endless, then the background of the sky would present us an uniform luminosity, like that displayed by the Galaxy—since there could be absolutely no point, in all that background, at which would not exist a star. The only mode, therefore, in which ... we could comprehend the voids which our telescopes found in innumerable directions, would be by supposing the distance of the invisible background so immense that no ray from it has yet been able to reach us at all.

If stars came into existence at a time t in the past, then we cannot see them at distances greater than ct where c is the finite speed of light (dashed line). This fact seems to have been first glimpsed by E. A. Poe (1848) and then stated clearly by J.H. Mädler (1858).

But this alone cannot completely solve the problem, also we need to add the condition that the age of the stars are limited.

Including Mädler and Poe, people lost interest in the debate on this solution for up to a century. When Mädler 15 years later in the review of the history of astronomy in his book commented the theory by himself that only the limited time model is "another possible solution."

Nevertheless, Overduin and Wesson (2009) still considered that limited time model correctly answered Olbers' Paradox, it is a more reasonable solution than Big Bang model. They used modern data to prove it again in their writing.

But if considering the "two suns problem," we know this is not a good model to solve Olbers' Paradox.

STATUS QUO OF OLBERS' PARADOX

By 1970, there had been about two dozen theories to explain the dark night sky, but only two were still active in academic circles: expansion of the universe theory and the finite age of galaxies theory (or, say, finite galaxy-emitted energy theory). The debate for these two theories focused on which was more correct and on which more accurately calculated the light intensity arriving at the Milky Way from extragalactic sources.

In this debate, the primary data used to support either theory was the extragalactic background light (EBL).

However, in the process of using the extragalactic background light to explain Olbers' paradox, astronomers committed a very obvious mistake, and their arguments become of little value.

From some of the latest literature on solving Olbers' paradox, such as "The Light Dark Universe" 2012 book by Overduin and Wesson, the extragalactic background light value used was $1.4 * 10^{-4}$ erg.

The extragalactic background light value that Wesson used in a paper in 1991 was 0.7 erg, much bigger than what he used in the 2012 book.

The data value they use is the result of a large number of observations and complex calculations obtained by the joint efforts of the entire astronomical community. From these value changes, you can see the efforts of the astronomical community over the past 20 years.

Unfortunately, despite the collective efforts of this astronomical community over the last several decades, the results have been getting away from the correct track, and have been getting worse and worse.

When looked at from another point of view, it is a simple matter to verify that the data is too small and getting smaller and smaller, more and more incorrect.

NASA data tells us that there are about 170 billion galaxies within the observable universe. Even if we only take 100 billion to calculate, each observed galaxy must have been able to send at least one photon to an observer on Earth. Using this data, and in accordance with the average brightness of each photon, the calculated result is 33.1 erg, which is much bigger than $1.4 * 10^{-4}$ ergs.

The brightness of extragalactic background light used in Wessen's paper of 1991 was 0.7 ergs, much bigger than the data he used in his 2012 book. The 1991 value was more reasonable. Therefore, after nearly 20 years of research on extragalactic background light, the astronomy community has been moving farther and farther from the facts.

I believe that none of the historical solutions to Olbers' Paradox have solved it correctly because they didn't carefully consider the light sources closest to the observation instruments. With the "two suns" hypothesis to test, no historical solution is compelling.

Is it possible that not use the Big Bang theory and other assumptions, in particular, not care whether the universe is infinite or expanding, whether the stars are uniformly distributed, or whether the propagation time of light is limited, but can still solve Olbers' paradox?

We will work on that idea next.

The Resulting Solution from the Historical Revelation - The Relationship of Bright Space and Dark Space

By understanding the various historical solutions discussed above and, at the same time, by contrasting them with the "two suns problem," we can see where the crux of the problem lies.

Whatever the hypothesis, no matter what the model, if the point of observation is between two suns, then there will always be a bright sky.

Therefore, the situation close to the location of the observer's space cannot be ignored. And that is precisely what all of the historical solutions ignored!

Intuitively speaking, the Big Bang, limited time, and other historical solutions have all come out of a huge problem: the solutions to Olbers' paradox seemed to imply that the universe should be dark. The question then becomes for these models: Where is the light? What is the relationship between light and darkness?

Please note that the history of the solution does not address the basic point: the stars near the perimeter of the space are certainly bright. So, dark space exists from where to where? How do you go from dark space to bright space?

This is very interesting.

The Sun is bright, and the space around the Sun is bright. Proof of this exists in all discussions on Olbers' paradox, but few talk about it. However, even far from a luminous celestial body, its surrounding space must be bright. What we need to determine is this: How big is this bright space that is brightened by luminous celestial objects? This also lets us know where the dark space begins.

In universal space, there are countless luminous objects. Therefore, there should be a lot of bright space.

If Olbers' paradox is correct, then all bright spaces need to link together to exclude the dark.

But, in fact, Olbers' paradox is wrong; the universe can only consist of alternating light space and dark space. Contiguous dark space prevents the presence of light in some areas, and the light cannot eliminate all dark space.

Therefore, to prove that Olbers' paradox is wrong, a theory would have to prove that the space between the countless luminous celestial objects consist of the light that surrounds each, along with dark space between them.

So, to prove Olbers' Paradox, we must consider the relationship between dark spaces and bright spaces.

And this vital issue, all historical solutions, including the Big Bang and the limited time available solutions have not fully discussed. There is no such model related to this vital issue in any historical solution.

Even the concept of the "dark" of the sky has no proper quantitative model to describe it.

Our study will be the first to take into account the local space of the observer (here, of course, the space around the Earth and the Sun), combined with the specific situation inside and outside the Milky Way, to explore why the night sky is dark. Its conclusions will then go from special inductive reasoning to a generally applicable method for the whole universe.

Before doing this, let's look at another quantitative (with some qualitative) analysis to determine the flaw of Olbers' Shell Layer model.

FLAWS OF OLBERS' SHELL LAYER MODEL – NEW SOLUTION BY DISTANCE

This is a "one light-year" example to explain how important the distance concept is, how carefully we should be using distance in the Shell Layer model, and how critical the role is of the "distance concept" in resolving the Olbers' paradox when understand from a view point entirely different from that of Olbers. After we understand the role of distance from the right perspective, then let's step by step compute comprehensive solutions to Olbers' Paradox.

Use the simple algebraic equation (2) from the introduction to calculate the light brightness intensity emitted from stars after a certain distance of propagation outside the sphere per unit area. The results, roughly calculated, reveal the large vulnerability of Olbers' Shell Layer model. Though seemingly perfect, the shell layer model, in fact, for most planets like Earth or galaxies like the Milky Way, basically does not hold.

The Shell Layer model is not true, of course, nor the establishment of Olbers' paradox. I am really surprised by the results. Let's see the details.

According to formula (2), the average luminance that the Sun sheds on the earth can be calculated as follows:

$$\text{Average light intensity on the Earth from the Sun} = \frac{\text{Light intensity of the Sun}}{4\pi (\text{Distance from Earth to Sun})^2}$$

Now I'm giving out a classroom exercise, one that requires the use of NASA's published data and the formula above to calculate: how far away from the Sun would the Earth have to be in order for the sunlight received per square centimeter per second to be only one photon?

The data NASA released can be found on the Internet. The average brightness of the Sun on Earth is $3.846 * 10^{-26}$ joules per centimeter square, while the average energy per photon is $3.32 * 10^{-19}$ joules, so you can figure out that the distance between the Sun and the Earth is about 1,010,000 light years.

That is, once sunlight shines through space for about 1.01 million light-years, at this distance in space, there is only one photon emitted from the Sun per square centimeter per second. Not so little photons the telescope is able to capture, the Sun disappeared in that distance. If you feel that this data is not small enough, the telescope still can receive photons from the Sun, then continue to increase the distance and continue the count. How far away will there be only one photon from the Sun per square kilometer per second? Can this photon be captured by a telescope with a receiving screen that is only a few meters in diameter? What is the noise from the universe that the telescope receives at the same time?

The cosmic dimension NASA has given is 13.7 billion light-years. In comparison, sunlight is spreading only in a space of less than a few billionths of the universe.

Let's come back to the topic of one light-year, there is another important calculation problem with regard to the Sun.

Our planet Earth is now located at 490 light-seconds away from the Sun. Now imagine the Sun is moved to one light-year away from the Earth. Of course, the sunlight the Earth gets is much less. To get the same sunlight intensity on the Earth as before, we need to add many Suns at one light-year away from the Earth.

Once the combined sunlight from all those suns shoots over, it will have the same light intensity as before the Sun moved.

Question: If the Sun is moved to one light-year away from the Earth, to get the same amount of sunlight intensity at one light-year away as before, how many suns are needed to add to that location?

I assure you that you can't guess right!

You can make your guess number as big as possible. The big numbers make it difficult to guess!

In simple mathematics, the problem can be written this way: at a distance of 490 light-seconds from the Earth, the light intensity on the Earth from one sun is X. Now put that Sun one light-year away from the Earth; what is the number of suns required to add, at one light-year away from the Earth, so that all those suns can make the same light intensity as X?

You can use the familiar formula (2). The step by step calculation is as follows: first, input the distance from the Sun to the Earth at 490 light seconds, then the formula can be used to calculate Earth's surface sunlight per unit area; when the results of this calculated, put it back into the formula to calculate the number of suns needed once the distance becomes one light-year.

But this way to calculate is not that smart. If you calculated this way, even if the result is correct, you only get a score of sixty.

The smart way is to use ratio and proportion we learned in primary school. You don't even need to know the Earth's surface sunlight per unit area. As long as you know the original sunlight intensity and the distance, you can get the result. The answer calculated in this way is worth 100 points.

The result obtained from the calculation is unbelievable:

If the suns are one light-year away from the Earth, we need to have 3,994,037,357 of those suns, close to four billion, in order to ensure that the Earth's surface gets the same intensity of sunshine as the Sun now supplies!

Distance, even just a single light-year, is a great magician, ah!

According to the above simple calculation, we know how a single light-year would quickly erode away celestial light as it traveled!

Thus, the primary flaw of Olbers' Shell Layer model is exposed! The problem is that the stars are not uniformity distributed in the space around the Earth, if you want to keep the Earth's brightness at night the same as during the day, requires more suns than the total number of stars in the Milky Way. Estimate quickly:

1 light-year from Earth, there is one sun, you still need about 4 billion suns to make daylight;

2 light-years from Earth, there are 0 suns, you still need about 16 billion suns to make daylight;

3 light-years from Earth, there are 0 suns, you still need about 36 billion suns to make daylight;

4 light-years from Earth, there are five suns, you still need about 64 billion suns to make daylight;

5 light-years from Earth, there is one sun, you still need about 100 billion suns to make daylight;

......

The Milky Way has a total of about 200 billion stars; the brightness of the Sun is moderate. So the light emitted by stars altogether is only equivalent to about 2000 million suns. Thus, even with the light emitted from all the stars in the Milky Way, it is impossible to fill the natural moat of the dozens of light-years around the Earth!

That is, the combined brightness of all of the stars in the Milky Way would not as bright as the sunlight that illuminates the Earth even if they were only just a few light-years away.

Through such specific, local, meticulous calculations, Olbers' Shell Layer model, with layers of light-year divisions, basically cannot be established! That is, the assumption that the celestial objects are evenly distributed is wrong, Olbers' model cannot be established.

Using the same calculation method that Olbers used, I have roughly proved that Olbers' paradox is not established because of the huge distances involved and because the surrounding distribution of stars near Earth is not uniform, so the light from all stars does not illuminate the space surrounding Earth. This hypothesis proves totally unnecessary the expansion of the universe. Even if the universe is infinite, not expanding, such proof is established.

The proof is actually a direct negation of Olbers' Shell Layer model, which he used to create his paradox. Since the shell Layer model he used ignored the important role of small space, the dozen light-years around the Earth, it looks perfect, but it is completely not true after careful calculation.

Understanding the Nature of the Problem with Olbers' Paradox That Needs to Be Addressed and Solution Highlights

From the foregoing discussion, you may already know that the historical solutions of Olbers' paradox have two common characteristics.

First, they look too far and too huge. They are only concerned with the big scene between galaxies and ignore the small Sun next door – almost none of the preceding solutions considered the specific circumstances close to luminous stars. In trying to prove that the night sky is dark, they ignored the fact that, near each luminous star, there is no dark night sky.

Second, their calculations are too rough and do not pay attention to the more detailed one or a few light-years in their computing. They don't have a concept of how insurmountable a natural moat of one light-year is.

The story of two suns tells us that, at near range, adding one more sun will create a completely different conclusion.

But this is not true at the not-too-far distance of one light-year; there, adding billions of suns is useless!

This is actually the opposite criticism of the theories of expansion of the universe, or those that just focus on the big picture of the whole universe, regardless of the space and celestial objects close to the observer. Even if the universe is expanding, if one more sun exists and if the Earth is between the two suns, then there is no night on the Earth.

From above, we see that, while discussing Olbers' paradox, we first need to look at the distribution of stars close to the specific circumstances of the near space of the Earth. Extending this to the general case, we need to identify the position of the stars, galaxies, and specific observing instruments to make a specific analysis. There is no universal model.

In fact, celestial bodies are unevenly distributed throughout space. The naming of the star, galaxy, cluster... is divided based on the size of the distance between celestial objects.

Because Olbers actually stood on Earth when he made his paradox, to solve the problem, we first have to deal with a range of local issues that are close to the Earth.

Below are five key points to solving Olbers' Paradox:

1. Why, when you stand on the Earth at night, do you only see a black sky?

2. The night sky is black, but how "dark" is black? Is there one quantitative reasonable description of the "dark" of night that we all agree on?

3. If the appearance of night on Earth is due to the Sun going down, then what

kind of role does the Sun play in this "dark" characteristic of the night? How do other stars that have a brightness different from the Sun divide the light and dark areas of their surrounding space?

After thinking about these, we leave the local Earth to focus on the problems in the bigger and more distant space:

4. After sunlight is blocked, why can't the unobstructed starlight from inside the Milky Way combine on Earth to illuminate the night sky?

5. Why can't the light from all the other galaxies combine to illuminate the piece of dark sky above the Earth at night?

Historical solutions to Olbers' paradox programs basically did not discuss issues 1-4 above. But we have seen that issues 1-4 are the key to unlocking Olbers' Paradox and are indispensable.

Now we will answer, one by one, the five questions above.

Why, when we stand on Earth at night, we see a black sky? (1)

What is now viewed from the Earth in the night sky does not look any different than what most people think or draw. Of course, my painting is not as good as some of those painted by other people.

The sources of the light shining on Earth are divided into three types according to the distance. From near to far, these are: the light from the Sun; the sum of the light from stars inside the Milky Way galaxy except for the Sun; the sum of other light emitted by celestial bodies outside the Milky Way. The three sources can be used to approximately describe Figure 5.1.

In the Figure 5.1, one side of the Earth faces away from the Sun. Because the sunlight is blocked by the Earth, a small cone of dark space is made. If a person stands in the dark space, the light from the Sun does not directly, or indirectly through reflection, enter his eyes. Due to the limited number of photons entering his eyes, he sees the darkness of space, and his eyes receive photons emitted from stars and galaxies. The space around the Sun is like the figure below, illuminated within a certain distance from the Sun, then slowly becoming dim until darkness takes over.

Therefore, light space and dark space are staggered over a vaguely gradual space.

If you need to draw a clear dividing line between light and dark space, you must first determine the value of the "darkness."

From the Earth, we can see the Milky Way, except for the area of space surrounding the Sun, which consists of illuminated space reaching a certain distance from each star and the dark spaces between, where there is no day and night, only the eternal black background dotted with countless other luminous stars.

Outside the Milky Way are the extragalactic objects (galaxies, star clusters or nebulae, etc.) located in space that humans do not actually know very clearly. We see numerous large and small bright spot unevenly distributed in the black background space. It is due to the uneven distribution of such objects, that people have been able to define galaxies, nebulae, and other similar collective concepts by distance between celestial objects.

Fig.5.1 Dark space regional distribution diagram, Note that the dark area of the Earth still contains light from other stars of the Milky Way and other galaxies in the universe. When we stand in the dark areas of Earth, the lights from stars shine on our eyes.

The distance of space surrounding each star that can be called bright is not clear; whether or not there is any bright area surrounding the space of the Milky Way is also not very clear because, even if bright space is there, we cannot detect it. The distance is too far away, and bright areas without celestial objects are too dark as energy propagation is limited across a finite distance. No matter what fancy detectors we have, due to the noise of the universe, these bright areas basically cannot be detected.

For example, we say that the side of the Earth that faces the Sun during the day is bright, right? The other side of the Earth, whose back is to the Sun is called night. However, what is the farthest distance away from Earth that this time can be observed? Even with the best telescopes, Earth's day cannot be detected at a distance beyond one light-year.

Therefore, we must first solve the sun-related problem of light and dark space surrounding the Earth. If there is a correct answer as to why the night sky after sunset is dark, it will also answer Olbers' paradox. As for proof of light and dark spaces surrounding galaxies or other extragalactic objects, we can only rely on vague speculation. We can come up with an algorithm, but can only say that the reliability of the calculated result is not so high. We cannot extend the results obtained from the little observation data collected on Earth to the whole universe. (From the large structure point of view, satellites above the Earth can still be taken as being on the Earth.)

The complete solution to this problem will be completed in part (2) later.

How Can Quantitative Methods Be Used to Describe the "dark" of the Night?

In the history of research literature, there is no quantitative definition of "dark." But without a quantitative definition, how can we have unified standards to calculate or discuss the problem of "dark"? If there are no guidelines, is it not a blind determination?

We consider the following aspects to determine the "darkness" of a moonless night. Should it correspond to the numbers of brightness (Physical units which represent the "photon energy received per square centimeter per second")? The following text has mathematics; if you don't like that, just skip the math.

The first consideration is the surface temperature of the Earth because this energy comes from the Sun. To use it as a reference index for our "dark" is reasonable. The lowest surface temperature measured on Earth is 184 K, the highest is 330K. The difference between them is 146K. So, I exaggerated a little. I used the Earth's surface brightness of the brightest millionth to mean "dark." This is possible, right? From NASA, the light radiating from the Earth's surface is approximately 1367.6 W/m^2, or 137 ergs, which is represented by $I_{EARTH} = 137$. In addition, we should note that night on the Earth includes light shining from out of the Earth, from other stars of the Milky Way, and from other galaxies in the universe, exclude those blocked by Earth.

Second, we know that the brightness of Pluto is darker than Earth's night. Solar illumination on Pluto is about 890 ergs. (The unit is too long and will be omitted.) Therefore, 890 ergs can also be used as one "dark" reference index, represented by $I_{PLUTO} = 890$.

Third, we know that the Milky Way has about 100-400 billion stars. Assuming, on average, that stars irradiate 200 photons per second per square centimeter on the Earth's surface, and the average energy per photon is $3.31 * 10^{-12}$ erg, then the energy received by the Earth's surface is 132.4 ergs. This is another number that might describe "dark" night, represented by $I_{DARK} = 132.4$ erg.

Fourth, from some of the latest literature on Olbers' Paradox, such as the previously mentioned work of Wesson, we can use $I_{Wessen} = 0.7$ erg to represent "dark."

In summary, we have obtained a series of night "dark" data, sorted by energy amount as follows (unit: erg/sec-cm^2):

$I_{Wessen} = 0.7$, $I_{DARK} = 132.4$, $I_{EARTH} = 137$, $I_{PLUTO} = 890$

We can put together all of these data points in preparation of a fuzzy "dark" membership function application.

Fuzzy Math is a simple but very useful tool that enables us to evaluate things more reasonably. For example, age is very important in the selection of managers. Suppose promoting the middle-aged, if the candidate is older than 50, he will be rated high, otherwise low. Mr. Sam, who just turned 50 years old, but in all other aspects, more outstanding than Peter, who is several months younger than Mr. Sam but not 50 years old yet. Because of the "old" criteria in the rating, Mr. Peter has been promoted. This is obviously unreasonable. So now we make a fuzzy mathematical formula to evaluate "age": let 20 years old be "young" and have an evaluation value of 1, let 70 years old be "old" with an evaluation value of 0. Then, from 20 to 70, make a "is young" linear formula as the fuzzy membership function of "age" and use it to calculate a person's "is young" level. In this way, the evaluation value difference between Mr. Sam and Mr. Peter is only a tiny few tenths, allowing a more comprehensive and scientific assessment.

The general meaning of the fuzzy membership function "dark" that we prepared is that there are different views of a value evaluation on "dark." The goal is to give different ideas of "dark" an evaluation. The evaluation comments are "absolutely dark, very dark, dark, slightly dark, bright." We use the formula to link data and in following table 5.1.

The score value in the table is calculated by interpolation of the formula.

The summary table of all aspects of integrated computing "dark" fuzzy evaluation formula is as below.

Table 5.1 "dark" fuzzy membership function evaluation

"Dark," description of reviews	Evaluation score calculation	Condition
Absolute Dark	1	brightness < I_{Wessen}
Very Dark	$1 - \dfrac{0.1\,(I - I_{Wessen})}{I_{Dark} - I_{Wessen}}$	I_{Wessen} < brightness < $I_{MILKY\text{-}WAY}$
Dark	$0.9 - \dfrac{0.4\,(I - I_{Dark})}{I_{Pluto} - I_{Dark}}$	$I_{MILKY\text{-}WAY}$ < brightness < I_{EARTH}
Little Dark	$\dfrac{0.5\,(I - I_{Earth})}{I_{Pluto} - I_{Earth}}$	I_{EARTH} < brightness < I_{PLUTO}
Bright	0	I_{PLUTO} < brightness

$$F_{Dark}(\text{Comment}) = \begin{cases} \text{Absolute Dark} & (1) & \text{if } I \leq I_{WESSON} \\ \text{Very Dark} & \left(1 - \dfrac{0.1(I - I_{WESSON})}{I_{DARK} - I_{WESSON}}\right) & \text{if } I_{WESSON} \leq I < I_{DARK} \\ \text{Dark} & \left(0.9 - \dfrac{0.4(I - I_{DARK})}{I_{PLUTO} - I_{DARK}}\right) & \text{if } I_{DARK} \leq I < I_{PLUTO} \\ \text{Little Dark} & \left(\dfrac{0.5(I - I_{EARTH})}{I_{PLUTO} - I_{EARTH}}\right) & \text{if } I_{PLUTO} \leq I < I_{EARTH} \\ \text{Bright} & (0) & \text{if } I_{EARTH} \leq I \end{cases}$$

Thus, the measured brightness is substituted into the formula, and we can determine if the measured sky is black or not black.

GENERAL ALGORITHM TO CALCULATE NIGHT "DARK"

Below we give an ideal algorithm to evaluate the "dark" at any point within the computing space.

Step 1: Choose the "black" values, such as the F_{Dark} (black);

Step 2: Select the point where the observer is located. Usually it is the Earth. In satellite running space above the Earth can be taken as on the Earth because the distance between the Earth and the observer is negligible;

Step 3: Choose a star that is closest to the observer. Calculate the brightness of the surrounding spatial extent of the star.

Step 4: One by one, calculate the brightness of the star and its surrounding area to see if there is intersection between this area and the brightness areas of neighbor stars. If there is, merge the two intersection areas into one and recalculate the brightness of this area; otherwise, continue to calculate the next star.

Step 5: Repeat steps 3 and 4 until all the objects within the space are calculated.

Follow these steps, first calculating the inside of galaxies and then, one by one, the clusters...

This idealized calculation is possible with the computational technology of today; you just have to put into a lot of resources.

Computing the bright and dark areas of the Sun in its surrounding space

If we choose F_{Dark} (black), the radius of the Sun's bright space is 49,900 light-seconds, the equivalent of light travailing 13 hours, 51 minutes, and 40 seconds. This is a big difference from 1 light-year.

That is, in the absence of consideration of lights that come from other sources, the Sun can illuminate a space with a radius of 49,900 light-seconds. We can also calculate the radius of the Sun's neighbor star Centaur Proxima Centauri C. In between this star's radius and our own Sun's, is a world of darkness. Figure 5.2 shows this situation.

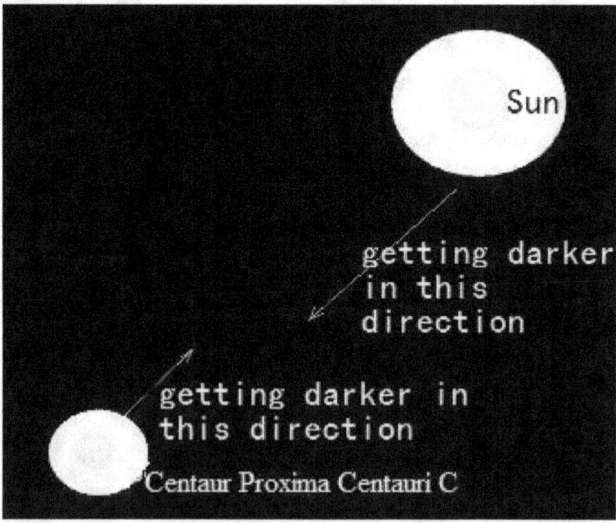

Figure 5.2 Brightness and darkness region of the Sun and Centaur C distribution diagram.

To estimate the light intensity irradiation from all the stars within the Milky Way to Earth

We first look at the emitting data of the stars near the Sun.

Table 5.2 The distribution of light intensity from all stars near the Sun

Distance to Earth Light-year	Star Number	Light intensity Solar unit
4	3	101.0101
5	1	0.01
6	0	NA
7	1	0.0001
8	5	100.03
9	1	0.01
10	4	1.0201
11	17	54.0604
12	7	1.0203
13	7	0.0402
14	11	0.0407
15	8	1.0403
16	3	0.0102
Summary	68	258.2924

Because the calculation is for the stellar illumination inside the Milky Way to the Earth, we use the old Shell Layer model to calculate. Note that in this calculation the sunlight is blocked by the Earth; almost half the light from the celestial objects of the universe is also blocked out. We didn't include them into calculation below.

Let's take a light-year as the thickness of the shell layer in our calculation. Although we already know that there are no stars on the second layer and the third layer, and only three stars instead of sixteen on the fourth layer, we still take the thickness of a layer of one light-year to calculate. Choose F_{Dark} (black) = I_{DARK} as the value of "black" of the night, then:

Since the Sun is in the first layer, the light intensity of the first layer can be regarded as $0.24 * 10^{-5}$ I_{DARK}.

The Shell Layer model assumes that, in the Milky Way, the light intensity on any one of the layers with a light-year thickness has the same light intensity as the first layer, which is $0.24 * 10^{-5}$ I_{DARK}.

According to NASA data, the Milky Way has a diameter of 10^5 light-years.

Because each layer has a thickness of one light-year, there are 10^5 layers in the Milky Way. So the total light intensity irradiated from all luminous stars of the Milky Way on the Earth is $0.24 I_{DARK}$, about a quarter of F_{Dark} (black). But do not forget that the Earth is blocking the Sun, and also blocking the starlight from that direction of the Milky Way. The blocked area should be more than one quarter of the Milky Way.

Therefore, the total light intensity of the Milky Way actually irradiated onto the Earth is also significantly lower than $0.24 I_{DARK}$.

ESTIMATE OF THE TOTAL LIGHT INTENSITY FROM EXTRAGALACTIC STELLAR IRRADIATION ON THE EARTH

Due to the great distance between galaxies, humans do not have much ability to actually spy on the secrets of other galaxies. According to data provided by NASA, the average scale of galaxies is 65,000 light years. Take this data as a layer thickness in the Shell Layer model calculation. We live in the Milky Way, so take the Milky Way as the first layer. Its central light intensity can be estimated like this: there are 1000-4000 million stars in the Milky Way, we will take its upper limit. Because of the moderate brightness of the Sun, we can say that every star has the average brightness of the Sun, and the average luminance level on each layer of the housing sphere will be $6.7 * 10^{-6}$ erg s^{-1} cm^{-2}. NASA tells us that the size of the observable universe is about 13.7 billion light-years, so the light intensity from all celestial objects of the universe shining into the Milky Way can be regarded as approximately $0.01\ I_{DARK}$.

So, what about the light from the stars beyond the observable universe? (If the universe was not formed in the Big Bang, and if outside the observable universe there is also unobservable space, or if the universe is infinite as Olbers thought, we can also roughly estimate this light.)

Using the Shell Layer model to the brightness of the Sun, the Milky Way, and the observable universe, we've discovered that as the distance increases, even if the layer thickness increases significantly, such as when calculating the total space volume of brightness from a single galaxy to the entire observable universe, the star brightness exposure to the Earth decreases rapidly.

At the same time, we must not forget that around the Earth there are more insurmountable moats that we need to use massive suns to fill. Refer to Table 1 above; if a shell layer is one light-year, then in the vicinity of 16 light years from the Earth, many layers need more stars to obtain the average layer light intensity. For example, there are no stars in the 6th layer, but in accordance with the uniform distribution of the Shell Layer model, this layer should have 36 suns; the number of suns in the 16th layer should be 256, but actually there are only 3. The distance between the stars of the Milky Way is relatively large, so on almost every layer, the actual number of stars in the layer will be a lot less than the number theoretical assumption requires for the uniform distribution of stars of shell layer model. The real data calculated leaves us a lot of room that is sufficient to accept the weak light from the observable universe and beyond. This weak light gets weaker, until it is close to zero, as the distance increases.

Because they are more distant, the light from them is weaker. Even in the infinity of space, the limit of the attenuating starlight coming from the outside layer of the infinite observable universe will tend to be a very tiny, fixed number. This number cannot fill those layers near the Earth that have no stars or are missing stars and cannot increase the brightness of the night.

As to what this number is, I do not have enough data, so we can only talk about this in general terms. It is left to posterity to calculate this number, after the Big Bang smoke is cleared.

Why, when we stand on the Earth at night, we see a black sky? (2)

Now we can easily answer this question.

Because of the unevenly distributed light celestial objects, because of the huge distance attenuation of light irradiation to the Earth, all the light from the stars of the Milky Way that shine through to the Earth is very weak, only increasing the night sky brightness a very small amount. And the light emitted by celestial objects from the entire observable universe that shines through to the Earth is also extremely weak, only increasing the brightness of the night sky a little bit. Even if the universe is infinite, dividing the infinite universe by distance into unlimited layers, these photons from those celestial objects to the Earth go through infinite attenuation sequences and will end because the limit theory tends to a constant. Because the number of stars within a few light-years from the Earth is scarce, this limit is not sufficient to replenish the constant light intensity that even just a few light-years distance requires.

Therefore, whether the universe is finite or infinite, is expanding or not expanding, under a premise without any assumptions, we proved that the Earth's night sky must be black.

In fact, when we assumed I_{DARK} as the night sky "black" quantitative description, we included those lights from the stars of the Milky Way, the observable universe, and outside the observable universe.

Is Any Place in the Universe Able to See the Black Sky?

On this issue, we basically have a negative attitude. Perhaps inside some galaxies that are full of light throughout the whole galaxy. Let's have a look at some pictures from NASA, such as Figure 5.3 of a starburst galaxy. Is there a dark area inside it?

Figure 5.3 Hubble Space Telescope image of starburst activity in neighboring NGC 1569 core area

Another example is based on the trajectory of Sirius. People predict there could be a companion planet. So, if the planet is always in between Sirius and Sirius B, there is likely to be no night there.

Therefore, the answer to whether or not it is bright or dark inside any galaxy in the universe can be determined by specific calculation. With regard to this determination process and the specific calculation model, we have discussed it above and will not repeat.

Are all galaxies like the Milky Way inside, decorated in shades of black that dominated their space? This is a big question; humans do not have the ability to spy on the details of the boudoirs of most other galaxies because of the huge distances between galaxies.

If it is not very special, generally the brightness in the space between the galaxies can be calculated in the same way that the brightness area surrounding the Sun was calculated. In accordance with the surrounding area of our previous bright galaxies, outside of those bright areas, the space between is black.

Overall, regardless of whether a celestial light emitting region shines its lonely light on its own, or a number of bright celestial objects form an alliance from a certain distance away, if the light-emitting celestial objects no longer exist, over distance, the darkness will eventually replace the brightness.

Revelation － Blind Spot of One Light-Year

The previous calculations fully show how "distance" plays a key role in the solution of Olbers' paradox. The closer an object is to the Earth, the more significant a role it has for our observations. The area of space "nearby" is what we should be concerned with, but it has been a long-forgotten corner. NASA's eyesight always focus on far away mysteries.

Because there is only one Sun near to Earth, in the evening, after the sunlight is blocked by the Earth, only weak photons shine in our eyes, and night falls.

Around other luminous objects in the universe, the night might or might not exist. To determine this, we need to follow the specific calculation with our given math model and can't give a generic conclusion. But one thing can be sure: when within the appropriate distance around a star and there is no other light-emitting celestial body, then the night will always be present at a certain distance away from the star.

Olbers' Shell Layer model was too rough an imagination, blocked the way to solve the problem through the use of distance, which is, blocked the right way to resolve the paradox.

That Olbers officially created this paradox is an interesting and important academic issue, but at the same time, the fact that he blocked the way to the correct solution to the problem shows that it was really both his success and his failure.

The impact of the false negative effect of distance of the Shell Layer model is not only limited to Olbers' paradox. A closer look at the various astronomical works and papers show: astronomers do not think that "distance" is an important factor in observing the universe. Many conceptual problems use the word "distance" – what we mean here is to understand the concept of a natural moat of one light-year. This is something that can be solved, but in a roundabout way with a variety of other methods. I think that this concept of distance is subconsciously ignored, in fact, because of the stigma of Olbers' understanding of "distance"!

Distance – not billions of light-years distant, but the close proximity of a light-year – has become a blind spot for the right solution of Olbers' paradox!

When we correctly understand this blind spot, the Olbers' paradox will not make itself a paradox. Of course, Olbers' paradox will cease to become one of the main supporting pillars of the Big Bang model.

In general, almost any point in space has different light intensity. Even within a black box made of particular substance, such as made of steel. Inside the black box there cannot be absolutely no light space, as the box made of any molecule material, the box itself will absorb and distribute strong or weak light. Possible exception would be a box made from dark matter that we leave to later chapters to discuss.

Therefore, to talk about whether it is dark or not at a point in space. we must first determine the definition of "dark" and its numerical description. In this context down to calculate whether light shines on this point is enough make it "dark" or "bright". From the foregoing discussion, we already knew that if at a point needs keep as brightness as that of the day on Earth, then about 4 billion suns are required to put at one light-year away. And if it is ten light-years away, then it needs 400 billion to maintain this daily solar brightness.

In short, no matter how big the size of the universe, and regardless of whether or not the universe is expanding, from the perspective of people on Earth, Olbers' paradox can be reasonably cracked. On Earth there will always be an alternating day and night.

So the Earth is not only in the light of vitality, but also in the dark sweet rest. This endless transmission of light and dark, yin and yang, will alternate here forever.

For any light-emitting celestial body in universe, when you travel a certain distance (e.g. 10 or 100 light-years) away from its lit space, the dark will gradually replace the light.

The debate between light and dark is never ending, as they alternate with each other under the harmonic of distance, a coexistent sport!

Epilogue

"Oh, the stars are too far away from us."

The father closes the book *Debate of Light and Dark*.

"It is such a simple truth. How come those big cosmic scientists do not understand? No, there must be something wrong," the son said.

Really makes people tangled, ah!

CHAPTER 6

TRUE FACE OF HIDDEN CELESTIAL OBJECTS AND DARK MATTER[4]

> Give me give me a pair of mental perception, let me put the world see a clear plainly genuineness.
> —— from "See the Flower Hidden in the Fog" lyrics of Yan Su

What we are seeking for?

Seeking for a planet that is outside or inside the solar system and that we can truly migrate to;

Seeking for a possible alien base;

These two tasks are actually the same thing; we need to identify possible goals from impossible celestial bodies.

SOLUTION ONE OF SEEKING: EXPLORING THE POSSIBILITY OF HUMAN MIGRATION OUT OF THE EARTH

On the topic of migration out of the Earth, due to language differences in English and Chinese, there is a huge misunderstood between the Chinese people and people in English-speaking countries. Overall, when the Chinese media translates NASA's news reports, most of the translated reports contain errors, misleading the public perception, and is also suspected of misleading public opinion.

---

[4] Please refer to the appendix "Hidden Celestial Object (HCO)" and accompanying text Reference

The reason for this misunderstood is very weird: in English there is no original word that means "may migrate to" (although, if there is, please forgive my poor English). I do not know if it is intentional or unintentional, but NASA has been using words such as "inhabitable."

So, regardless of whether NASA intends to say "may migrate to," the English news reports only use such words as "habitable" or "inhabitable." And these two words mean "human beings can live there" rather than that people are able to move to there to live.

NASA, year after year, publishes news that a new "habitable" planet has been discovered. Such a description has been misleading, and people mistakenly thought it was "possible to migrate to" the planets NASA discovered, but NASA did not directly say "people can migrate" there.

As more Chinese news reports take NASA's vague hidden meaning as the announcement, let's look at the following news about the discovered planet Glises 561. Figure 6.1 shows one of the numerous similar reports of the search results obtained in Baidu Online with the key words "外星移民 alien immigrants." We can see in the figure that the report embellished and misinterpreted "habitable planet" to mean that "migration to the planet is not a dream" – that is, the dream of human migration out to the planet is no longer just a dream, but rather, migration to the planet can be really achieved.

Today (2015-6-27), a query on Baidu search engine with Chinese keywords "可移民星球 (be able to immigrate to star) "got 3,180,000 pieces of related content.

But if we want to check on Google the same meaning in English, we will be helpless: there is no way to enter the word "可移民 (be able to immigrate to)" in English. The word "immigrationable" does not exist. Only with "habitant" or "inhabitant" keywords that we can find associated news from NASA.

The exaggerate reports of these kind of scientific and technological achievements is not that wrong, but if the point is completely misunderstood, and the consequent misleading of the general public, which should still be reviewed. Obviously, it's just an elusive moon cake, but is exaggerated into a delicious dish feast on the table. Though I cannot completely blame them, this is really caused by NASA repeatedly announcing the discovery of habitable planets and by the translation difficulty between the two languages on the topic, and thus we get these kinds of scripts.

In recent years, cosmologists have put out such numerous reports and made such an ambiguous atmosphere, along with the vague media reports, that all the ordinary people who are not scientists truly believe that cosmologists are able to lead us in our time of need to successfully flee from the Earth. I believed it too.

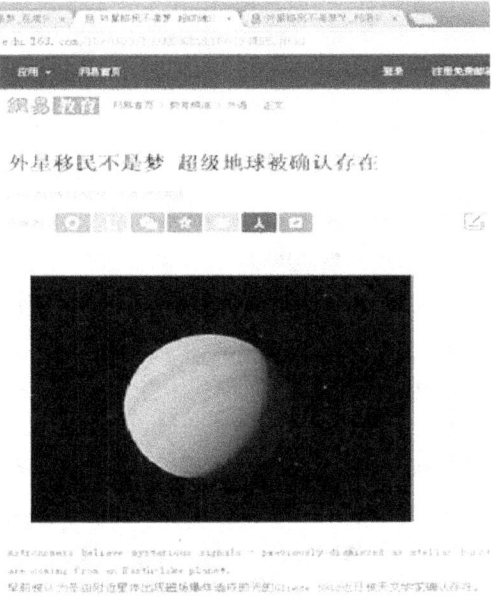

Figure 6.1 Chinese media reports that migration from Earth is not a dream

Migration off the Earth is not a dream, ah!

But, in fact, it really was, and still is, a pure dream!

And in the future, even if you count future years in ten thousand year units, it would still be a pure dream – if we continue in the current way of thinking and insist that the current cosmological theory about the universe is correct.

Brutal fact:

So far, outside of solar system, humans have not found a single planet that we are truly able to migrate to!

According to existing cosmological knowledge and theory, there are also no prospects to be able to make it true.

So, dear readers, be patient and go through our theory to see the light of hope.

Minimum conditions of a planet that human beings can migrant to

If humans want to migrate to a particular place from the Earth, it must meet at least two conditions:

1). It must be habitable.

2). It must be reachable.

To immigrate to a planet, it first needs to have a habitable condition. If not, of course, we will not immigrate there. If we immigrate to a star, like the Sun, is it not a rush to commit suicide?

In accordance with the standards of human needs on the Earth, cosmologists have found many planets that humans can live on. The closest ones are only about five light-years from Earth, and the ones far away are more than two thousand light-years from Earth. There are countless further away, but cosmologists no longer overemphasize them since they are actually aware of the distance problem.

The problem is that the planet must meet the second immigration requirement: must be able to reach it.

Cosmologists usually talk about billions of light-years, painting maps of the sky that span billions of light-years, and talking about the scientific story of things occurring billions of light-years away. Their lofty eyesight ignores things just a few light-years away. One story the book will discuss later is how cosmologists reporting on Abell 754 said: "Abell 754 is relatively close to Earth, about 800 million light-years away."

Eight hundred million light-years, ah? "Relatively near the Earth?" How ambitious are their minds?

Unfortunately, I'm just a little person who is only concerned about the immediate reality that occurs within a small distance from me.

I just wanted to figure out: how far is one light-year?

Cosmologists have told me that asteroids might hit the Earth, or that the Earth will be destroyed in the future. But do not worry, cosmologists have found and given us a good way out. April 18, 2013, NASA again announced its discovery of three "new earths." In addition to the Earth, in the vast universe, there are also Earth-like habitable planets. The news said that the international media is very excited. Kepler Space Telescope has observed two extra solar planets so far that are "the most like Earth" and that "may be the most suitable for human habitation." The planets are "moist, warm like Hawaii, cold like Alaska," and 1200 light-years away from the Earth.

Being a coward, I have to prepare good to approach there. Let me first calculate how much food I'll have to prepare for the long road to that planet.

Such a calculation, Wow! My eyes were sent straight!

Guess how long it takes to fly there, OK?

Currently, riding on the fastest spacecraft of NASA, a distance of one light-year

Chapter 6 True Face of Hidden Celestial Objects and Dark Matter

requires thirty-seven thousand five centuries to complete!

To get to the nearest habitable planet five light-years away, the flight time is one hundred eighty-seven thousand five hundred years!

To get to the newly discovered "moist and warm like Hawaii" habitable planet, it will take forty-five million years!

Ten thousand years as a unit, yeah!

If someone said to me, "I'm sure you'll be the richest man in the world after 1000 years," I wouldn't be very thankful and, perhaps, wouldn't be able to help punching him in his big mouth.

Does it make sense to repeat talking about what might happen ten thousand years from now?

Anyway, we have to remember the following brutal fact: so far, outside the solar system, humanity has not yet found a planet that meets the two basic conditions above.

Migration out of the solar system is a dream that humans need to implement as soon as possible, but still have no hope of realizing

The tragedy is that mankind has had to face the reality of issues that need us to be able to migrate to planets outside our solar system, which is a big problem related to whether human beings can exist long term or not.

Figure 6.2 is a screen shot from Netease news. In the title, "Immigration from the Earth is not a dream?" the reader needs to pay special attention to the question mark. This is different from Figure 6.1, which, in a positive tone, tells people that "immigration from Earth is not a dream!" The two are completely different. The question mark points to the extreme amount of time it would take, in ten-thousand-year units, for humans to migrate to an extra solar planet, which, in fact, denies the basic possibility.

The illustrated article discusses the following issues with extra solar immigration:

1. Relocating to an extra solar planet has been the dream of mankind.

The famous astrophysicist Stephen Hawking has pointed out that mankind must migrate out the solar system to escape the fate of extinction. An asteroid impact on the Earth, or the threat of the nuclear war, sooner or later will destroy humanity. The solar system is too close to the Earth, so humans must find habitable planets outside the solar system.

173

National Aeronautics and Space Administration (NASA) Administrator Michael Griffin has said that, in theory, the planets species are unlikely to survive forever. He told us: We have strong evidence that, on average, every 30 million years, the Earth's species will face a large-scale destructive event. One day, we must immigrate to alien world, but I do not know when that day will be.

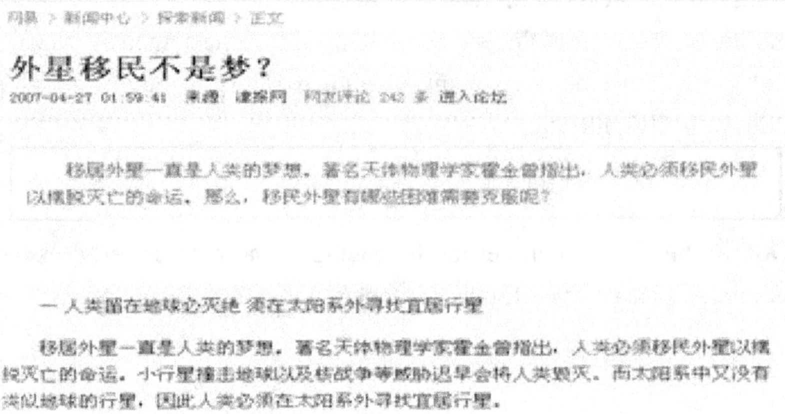

Figure 6.2 Immigration from Earth is not a dream?

2. What are the barriers to be overcome when migrating from Earth?

The first hurdle is to establish livable standards for humans and to find those planets that meet those conditions.

The second hurdle is to reach the livable planet. If you follow the current speed of the space shuttle, to get to a planet about 4 light-years away from Earth, we will need about 15 million years for the trip. To be able to migrate to an alien world, humans must create vehicles that can travel as fast as the speed of light. The reporting Figure 6.2 said: "It is reported that the US Air Force and NASA are currently working on a secret research 'spatial engine' aircraft, and are expected to create a prototype for testing within five years. If the test is successful, then it would be able to travel from Earth to Mars in only 3 hours, and go to a planet 11 light-years from Earth in just 80 days."(That is to say, the vehicle speed is 50 times the speed of light. Brag a little spectrum, OK?)

The third hurdle is that, after immigrating to the alien planet, mankind now has to solve the issue of how to live there. Currently, the United States and Russia and other countries have cultivated more than 100 kinds of crops, fruit flies, spiders, fish, and other animals in weightlessness on the International Space Station. If this

Chapter 6 True Face of Hidden Celestial Objects and Dark Matter

technology can be applied to the livable planet, human problems of survival can be easily solved. In addition, the ability of humans to survive this immigration is a problem. A French scientist discovered that, in weightlessness, important structures of living cells cannot properly form, which means that humans cannot live and thrive long in near weightlessness.

See the above description; what impression does it give you? Do you not feel that "immigrant to an alien planet is not a dream," without a question mark?

However, the fact is, even after thousands of years, that this is very likely an unattainable dream!

The second hurdle was described in a very disgusting "It is reported that" it will only take five years to create prototype that can travel as fast as the speed of light? "Go to a planet 11 light-years from Earth in just 80 days"? If one can run as fast as light, he needs 11 years to arrive 11 light-years away. Is there any basic common sense or knowledge?

The illustrated report in Fig.6.2 was issued in April 2007. It is now in June 2015, eight years later, and the current speed of US spacecraft is almost the same as eight years ago, no different. From a technical point of view, based on recent scientific progress, the idea that we will soon "Create a vehicle as fast as the speed of light" is just impossible, only a dream! To be precise, it is impossible for anything to travel at the speed of light, because, according to Einstein's mass-energy conversion law of matter into energy, at the speed of light, that thing will no longer exist!

Of course, with technological advances, the speed of human spacecraft will be improved. But this will leads conflict with Relativity theory. This is something we leave for later specific discussions.

The article also says: based on our current situation, humanity's ability to immigrate to an alien world seems hopeless, but can we change the way of our thinking? Dr. O'Neill at Princeton University in the United States believes that the best approach is to build a city in space so that all human can relocate to the city to live in space. Hundreds or even thousands of years later, the atmosphere of Earth will again become favorable, in the case that the Earth is not damaged by humans, and humans can return to Earth later at that time.

Please note the sentence of great truth that is flooded by large sections of text: "Based on our current situation, humanity's ability to immigrate to an alien world seems hopeless."

This is the cruel truth we are facing: humanity has an urgent need to immigrate to extra solar planets, but human beings have yet to find a planet we can truly immigrate to outside our solar system.

Where is the planet that humans might possible migrating to?

Cosmology should not be giving a boundless hope first yet then get misty on tens of thousands of years and not clearly tell people that it's just a picture to be drawn for thousands of thousands years.

According to current cosmic theory, the cosmological theories, and the scientific development of speed, mankind has no hope of migrating to extra solar planets.

However, Heaven must be a way.

If we change our way of thinking, broaden our focus on the eyes of the universe, and update our theory of the universe, we will find true extra solar, livable planets that we can migrate to. It is possible that we can reach a place in the extra solar planets!

This place is hidden in the celestial bodies that we are going to propose in later sections.

SOLUTION TWO TO FIND: IF ALIENS EXIST, WHERE ARE THEIR BASES?

We know some people do not believe in aliens. Others believe; some have said they met the aliens, and some have said they had dealings with aliens.

In California's capital city, Sacramento (三块馒头 San Kuai Man Tou), a group of people occasional meet who claim to have had dealings with aliens.

My close friend saw a very strange UFO but did not see aliens. I'll relay his story later.

Let us tentatively assume that aliens exist. Then there are questions that naturally follow:

Where are the alien bases or homes?

Alien aircraft take off from where?

Alien spacecraft from Earth fly back to where? They should have somewhere to go.

In fact, there are many issues involved, including the question of whether there are aliens. If no one can give a reasonable explanation of where the alien bases or homes are, then it indirectly denies the existence of aliens.

However, if we can give a reasonable explanation to the question of where the

Chapter 6 True Face of Hidden Celestial Objects and Dark Matter

alien bases are, it will greatly enhance the possibility of the existence of aliens.

Often, people will say: our telescopes, including space telescopes as powerful as the Hubble telescope, which can observe stars that are hundreds of millions of light-years away, have not found any alien bases, and therefore, alien bases do not exist; or aliens should only be present in very distant places, such as a hundred, thousand, million light-years away, so we cannot see them through the telescopes. The following discussion on the Hidden Celestial Objects will tell us that this is a completely wrong idea.

In ancient Egyptian pyramid texts, we often see a large number of astronomical stories that tell about aliens from distant places of the universe. For example, the Titans came to Earth after "millions of years" of travel in the dark with no air in the universe; the god of wisdom, Naxos, who counts the number of stars in space and measures the land on the ground, has a magic power that can make an already dead Pharaoh have several hundred years of life; the eternal god Osiris traveled for millions of years to get here. In addition, there are many puzzling in Scripture, like "millions of years" and "one hundred millions of millions of years," and so on. These are telling people that aliens in ancient Egypt came from distant planets.

However, from another point of view, if the aliens who frequently appear on the Earth come from faraway bases, then what kind of speed should the alien spacecraft travel at so that it can freely haunt the Earth?

We already know that a light-year away, with the existing fastest spacecraft on Earth, requires more than thirty thousand years completing the travel. If the alien base is hundreds of light-years away, even if the alien aircraft fly to Earth at the speed of light, one Earth visit would take hundreds of years, which is more than people's whole lifetime.

Some people might assume: alien life is very long; our one hundred years might be the equivalent of one alien year. This may be possible, but our reasoning does not base itself on such a myth. We make our assumptions with mankind as a reference. Suppose the basic physiological conditions on the alien planet are not that different than for humans on the Earth. So, a hundred years' time is relatively long.

Some people would say that the aliens come to the Earth through a black hole, or a wormhole, etc.

I hereby solemnly declare: this book does not use any of these seemingly "scientific" ideas, for which there is no scientific basis and which have never been proved with a scientific experiment, existing only in the theory of contemporary cosmologists, Hollywood star war blockbusters, or the fantasy fiction of so-called "theory." Such theories are distinctly different with the assumption of the possible existence of aliens.

Specific to this chapter, time travel through wormholes or black holes exists only in the imagination of theoretical cosmologists. There is no a convincing scientific experiment or observational facts to support these ideas. Even the proposed existence of wormholes or black holes itself is still not supported by solid scientific experiments of any observations but is a result of indirect bold speculation based on a number of observed phenomena, which we will discuss later in detail when talking about Hidden Celestial Objects.

However, the numerous eyewitness testimonies since ancient times of alien spacecraft may be said have a proven track record over the years. It can be said that the research on alien exists and there is a solid basis to making a scientific observation project for further work.

Everything this book describes or will describe is based on empirical scientific experiment or observation of the facts, not unfounded speculation and imagination. It is not a purely theoretical speculation, but a combination of theory and facts.

In fact, according to the existing theories of physics, aircraft flights at the speed of light do not exist because, at the speed of light, the material will be converted into energy, the aircraft and astronauts will disappear and turn into energy. In the later discussion of the theory of relativity, we say that "near light speed" is an ugly concept. The orthodox astronomical community does not agree that the speed of light can be changed or made slower. But when in need, they use the ugly concept to discuss a lot of topics.

So, if aliens really are in the distant reaches of space, it is almost equivalent to aliens being non-existent because they would take too long time for a trip to access the Earth. On Earth, distant alien spacecraft should only occasional visit, and each visit may be very long.

The history and current recorded facts are: alien spacecraft frequently haunt the Earth, and there are many different types of space aircraft. Some of these vehicles are common, such as a disc-shaped UFO; some are rare, like square or oval shuttle.

It is concluded that, not only there are occasional guests to Earth from very distant depths of space; more aliens are coming from our near neighbors.

That is, according to analysis on UFOs in human history, the conclusion is: if aliens exist, then some alien's bases are very far away from the Earth, but some, perhaps, are not far away and just in our neighborhood. Some may even be in the solar system, or near the solar system.

Chapter 6 True Face of Hidden Celestial Objects and Dark Matter

This is an argument that overthrows public and expert knowledge.

It is generally believed that, with telescopes, we basically have clearly seen the area of space very close to the Earth (such as within a few light-years or several tens of light-years); only very distant celestial space cannot possibly be seen by powerful instruments like the Hubble Space Telescope.

Unfortunately, this perception is totally wrong. Indeed, it is entirely possible that the most powerful human telescope cannot see celestial objects that are very close to us.

Fortunately, if we promptly recognize this, then it is possible to solve the problem of finding the alien bases. More importantly, in this new way of thinking, the idea in the introduction to our discussion that looking for a suitable planet for humans to inhabit another planet is not just a slim dream. Instead, it will become possible to achieve a great hope.

This chapter will clearly describe the big, popular articulation of this view; the specific mathematical proof is attached to the back of the book's appendix.

Through the process of searching for possible extraterrestrial bases, and the associated process of looking for a planet for humans to immigrate to, it can be seen that, with regard to the problems of the universe, even cosmology experts are also prone to cognitive bias. Correcting these deviations can bring great benefits to humans.

The estimated range in which possible alien bases exist

Let's start with a reasonable range estimate on where alien bases exist to further explore some of the problems about finding alien bases.

First, we can compare the historical record to the Earth's existing space technology, the pace of development of science and technology, and so on, to make a rough estimate of alien level of technology, for example, about how advanced the alien spaceship is and what its speed is.

From the comparison of sightings of alien spacecraft in history and currently, and from technological progress on the Earth, to speculate, the pace of the speed of the alien spacecraft is now probably a hundred times that of the fastest human spacecraft speeds, a thousand times or 10 thousand times; such a large range estimate, should be somewhat reasonable,

From the historical record, since a few thousand years ago, there has been records about alien flying saucers landing on Earth. But during that time, their technology was not so advanced, so their early aircraft would cause a huge ring, a loud sound, and radiant light.

The UFOs from reports in recent years present a different point of view, such that the alien flying saucer appearance, sound, and other aspects have been greatly improved. Like the UFO my friend witnessed, the shape has become smaller, and there is almost no sound when flying.

There are many historical records. In China, from the Qin Dynasty to the present, various uninterrupted records report on unidentified objects that came to the Earth. I have excerpted the following:

The Jin dynasty mythical masterpiece "Notes of Pickup Oblivion" Volume of Tang Yao, in which there is the following account: After Yao became Emperor for thirty years, a huge ship appeared in the West. There was a light at night on board, and the people at seaside called it "贯月槎 Ship Go through Moon." On board were flying fairies wearing white feathers. This story happened in 2327 BC, dating back 4,000 years ago.

In the chronicle book "资治通鉴 Governor Mirror of History" written by Sima Guang in Song dynasty, there are 17 suspected alien astronomical records and related activities. For example: in the Jian-yuan second year of the Western Han Dynasty, in the summer in April, a star like the Sun came out at night. This is a record of the object as bright as the Sun that appeared in the sky at night.

Song scientist Shen Kuo's book "梦溪笔谈 Dream Creek Essays" also recorded UFO stories. In the Yangzhou region's lakes there was a strange big bead with a clam shell shape, a typical UFO shape. It emitted intense light, stayed there for more than ten years, and switched its residence between three lakes, many residents witnessed. This may be an account of visitors from distant space, where it is not easy to come from, so they stayed longer.

Figure 6.3 "赤焰腾空 Red Fire Soaring into the Sky" is the work of Qing Dynasty painter Youru Wu in his old age of 1892. Figure 6.3 is considered to be a detailed and vivid UFO sighting.

The screen diagram of "Red Fire Soaring into the Sky" depicts the scene when, on the Nanjing ZhuQue Bridge, pedestrians crowded like clouds to all look at the overhead sky, eager to watch a group of gleaming flames. In the inscription at the top of the screen, the painter wrote: "September 28, at eight o'clock in the evening, Jing Ling (now Nanjing) south, a group fire blankets the sky from west to east, shaped like a giant egg, red light without shining, floating through the air with very slow movement. At that time, the floating cloud hides the sky, the heaven is

Chapter 6 True Face of Hidden Celestial Objects and Dark Matter

darkened. For heads looking up, the sight is very clear. More than hundreds of people stand on tiptoe on Zhuque Bridge, looking for time upon the sight. With increasing distance the brightness decreases. Somebody said there was meteor in transit, but that should be fast and appear instantly. But this ball from near to distant, gradually disappeared, like stagnating water, so we knew that was not meteor. Some guessed that there were children putting lanterns into the sky, but at night, the storm was a north wind, and the ball turned east, so we know it was not sky lanterns. Testimonials guessed a lot, but have no way to speculate an accepted result. An old man said that, at the beginning, he could hear a weak sound that one had to be quiet to hear. It flew from outside of the south gate, giggled, and then carried on in a different direction!

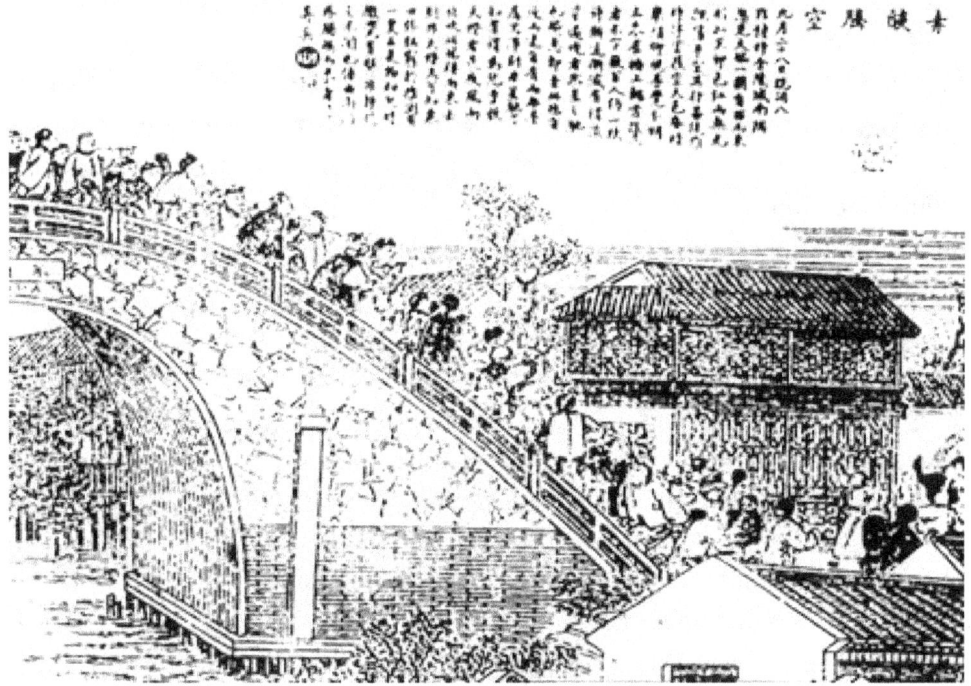

Figure 6.3 Chiyan vacated FIG.

In this detailed and vivid illustration of sightings in the city of Nanjing, the fireball passing time, location, number of people who witnessed it, the fireball size, color, light intensity, flight speed, all were in clear description. It is a qualified academic observational report.

In recent years, newspapers, magazines, television shows, and media networks have been filled with a steady stream of sightings and reports on alien encounters and alien spacecraft.

My friend told me that he saw a UFO like this:

It was in the evening of September 9, 1997 7:30 pm, on the way back from sent his son to a piano lesson, at a few hundred meters away from his house in front of the traffic lights, suddenly he saw in front of him about four or five hundred meters away on the roof of the houses at the foot of the mountain, a strong wind blow a newspaper shuffling along rows of roofs. My friends feel very strange: why the wind so strong tonight? Driving into the home parking lanes, he saw an object like a piece of newspaper staying in front of the house about 8 meters above the poles. Its body was around a slight yellow light flow, making the material looks more like a blow by the dark side Yellow flag He realized that he was encountering a UFO and was shocked, and hurried back to the house to get his camera. Meanwhile, the UFO had transferred from the tile into a flag-like shape. It quietly rose into the sky and flew off to the west. While its light was not bright enough to be captured by the camera, after a long time in the air, it could still be resolved from other stars in the clear sky.

The UFO my friend saw was both small and silent. If you assume that an existing spaceship on Earth has a similar shape and speed as an alien UFO from thousands of years ago, the idea that now, after thousands of years of evolution, the aircraft is a hundred thousand times faster is a logical inference.

Using this rough estimate to measure vehicle speed of those UFOs that frequently appear on the Earth, the alien base should be near the Earth, from less than a light-year to just tens of light-years away. According to our estimation of possible alien spacecraft speed, the aliens would be able, with no difficulty, to depart from the base and frequently visit the Earth.

Their base might also be not very near the Earth. Otherwise, aliens would take Earth as their roots and often come in droves.

The Dogon people of Mali, Africa, are still in a primitive tribal society. But the tribe has the legend that they are from the third star of Sirius, a planet of Sirius. The planet is covered with water and is 8-9 light-years from the Earth. This is our estimate of the alien base location that is relatively close. So there may be a base or or a hometown for aliens.

If this alien spacecraft is from a place that is not too far away, why hasn't the powerful Hubble Space Telescope found the planet where the alien base is? Why hasn't it found the ancestral home of the Dogon people near Sirius?

With respect to the broader point, might there exist places near the Earth that

Chapter 6 True Face of Hidden Celestial Objects and Dark Matter

even the Hubble Space Telescope cannot see?

Yes.

Such kind of planets are indeed present.

And this kind of planets that even the most powerful telescopes cannot see. There is a considerable amount of them, both near and far. In each local area in space, the number may be unimaginable.

This may overthrow the concept people have inherited. But this is no doubt fact, and we will prove it below.

And people do not see them, mainly because the vast majority of them really cannot be seen. There are others that we should be able to see but, for various reasons, have not found.

A few should be found but haven't been so far. One possible case for this is that the alien base is masked by aliens using stealth technology to shield it from our detection.

Another possibility is that, in the years after the invention of the telescope, people have slowly focused more and more attention on distant space, and the most advanced telescopes do not focus on the near space, so some objects in the vicinity that should have been found years ago still have been overlooked.

More likely is that those hidden objects, although indeed in our neighborhood, cannot be seen. They hide from our current telescopes' vision.

So, when we put the focus of the most powerful telescopes on near-Earth space, it is possible that we might find some close by, but we have not found such objects yet. In this regard, I have a very optimistic attitude because many of these objects may not be alien to us and may not be blocked from us. But due to our own reasons, we lose sight of our near neighbors.

So, inevitably, there are many celestial bodies that exist far and near but that have drifted away from our vision. Our research needs further approaches to find them.

I named these temporarily or permanently invisible objects "**Hidden Celestial Objects**" (**HCO**). Specific scientific proof description please refers to Appendix papers.

Those alien-infested bases near the Earth should be from some Hidden Celestial Objects relativity near the Earth.

So, what is a hidden celestial object?

HIDDEN CELESTIAL OBJECT OVERVIEW

Later, we'll give a precise definition; here is just a brief overview.

As the name suggests, it is an ordinary celestial object, but it exists outside of our vision, and we cannot see it.

However, we do not need brighter eyes; we just need to use our brains and a little bit of reasoning so that we may "see" them.

Hidden Celestial Objects are everywhere, from near the Earth to hundreds of billions of trillions of light-years away.

Hidden Celestial Objects are not like the mysterious Dark Matter that is still vague, without any specific example or experiment to confirm its existence. Instead, dark matter belongs to a small category of Hidden Celestial Objects.

The number of Hidden Celestial Objects is far more than the number of found objects.

Human eyes, even if aided by the power of any imaginable blessing of an advanced magic telescope, will still be blocked at the gate of truth by a few natural moat barriers included in the "distance."

To figure out the sources and applications of Hidden Celestial Objects has significant meaning. Through them, Hubble's law of expansion of the universe can be examined, and we will even be able to find the true source of the cosmic microwave background radiation, etc. These, in our subsequent chapters, will continue to be discussed.

With the hidden celestial object concept, our vision flies out of the little universe delineated by NASA and generated because of the Big Bang, flies through the fog of the Big Bang, and "sees" the true face of the broader magnificent universe.

Let's go through following sections to initially know about the true face of these Hidden Celestial Objects.

Comets whose cycles bring them near to Earth are typical Hidden Celestial Objects

For instance, the Halley's Comet that we have watched with our naked eyes for thousands of years, every time when it moves from the Earth, it first hides from us, and then comes into our eyes once it moves closer. Before it is observed, it is a hidden celestial object; later it is seen in the eyes of people, then it is not a hidden celestial object anymore; and when it leaves the Earth and goes away some distance, people's eyes no longer see it. Therefore, with respect to people's eyes, it becomes a hidden object again.

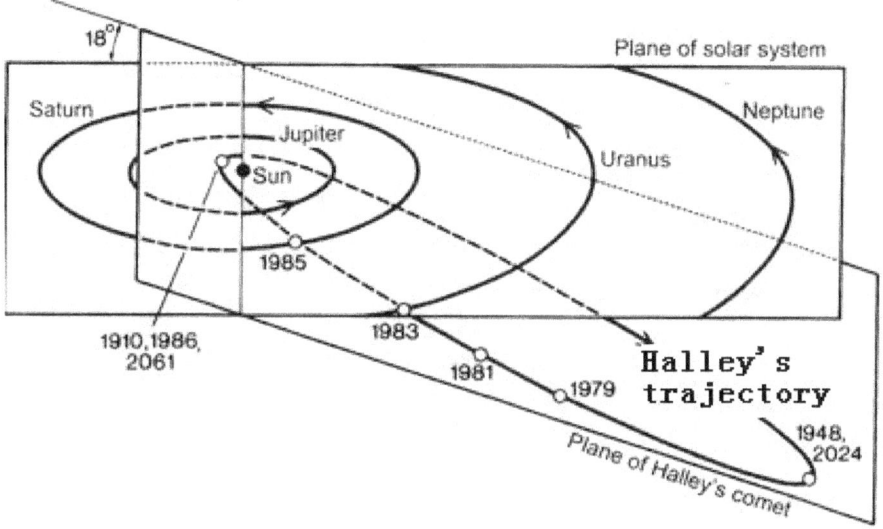

Figure 6.4 Halley's Comet track

This actually means that for an ordinary object, when conditions of viewing it change, it is likely to go from being visible to invisible, or from being seen to not being seen.

Watching Andromeda with the naked eye, it is a dark oval spot, weak and vague. An interesting phenomenon worth to notice is that our naked eye can see the Andromeda galaxy 2.54 million light-years away, but it cannot see Proxima Centauri 4.2 light-years away. Why? The common explanation is this: talking about how far an observation instrument can see is meaningless; it is better to say what brightness it can see. This explanation is incomplete.

We need to examine all the variations of related factors and to find out the law of the variation.

Watching the comet involves several aspects to be analyzed: the light received by the eyes, the light irradiating from the comet, and the distance between the eyes and the comet. Careful study of various possible changes among the three will allow us to find some very interesting patterns.

SEVERAL FUNDAMENTAL FACTORS AFFECTING THE ABILITY TO SEE CELESTIAL OBJECTS

This basic rule starts with the simple mathematical expression of formula (2) in our preliminary knowledge section. We copy it here:

$$\text{Light Observed} = \frac{\text{Light emitted from Celestial C}}{4\pi \text{ Distance } R^2} \qquad (2)$$

Substitution what we are to discuss of the subject into the above formula, we get:

$$\text{Light eyes received} = \frac{\text{Light emitted from Comet}}{4\pi \text{ Distance } R^2}$$

Seeing this equation, the first thing we notice is that the light eyes taken is inversely proportional to the square of the distance. In other words, it reduces very quickly with distance. According to what we discussed in the preparatory section on the limit law, the light our eyes see will finally be zero.

That is, at different distances, Halley light irradiation on our eyes is not the same. When the distance is big, the light from Halley received by our eyes is weak, and, in contrast, when the distance is small, the received light is strong. When the distance is far enough, the eyes can't sense the light.

Suppose Halley's Comet irradiated light intensity is essential constant: there is no big change. So, in this case, the important factor affecting whether or not the Halley's Comet can be seen at a certain distance is the observer, here it is the eyes.

Good eyes, of course, can see more clearly; there is no doubt.

Then, with the same pair of eyes, under what circumstances do we not see Halley's Comet light shine through?

Let's first understand how the eyes see things.

The observing limit of the eyes

There was a piece on a Wechat about an illiterate who encountered a group of Magi discussing why big rain in dead people. The magi cited formulas and reasoning until, finally, the illiterate said one thing: Who of you have not been

Chapter 6 True Face of Hidden Celestial Objects and Dark Matter

drenched in rain? So the circle suddenly deserted town. The illiterate was very proud that the magi were suckers.

Actually, the topic being discussed was, why can't the rain hit down flying mosquitoes?

Some questions may seem simple but really contain very profound truth. But if you do not stand on the height of the mountain, you cannot see the scenery at other side of the mountain.

Now here is a silly question: can eyes see a photon?

The problem here is actually divided into two aspects. First, can the eyes accept a photon?

The answer is obvious: yes. If the photon goes into the eyes, can the eyes refuse to accept it?

So, the eyes accept a photon. Now we will deal with the visual nervous system, can it handle the information of one photon? Will we eventually feel what it is to "see" the photons?

After a photon enters the eye, the sensor in the eye on the retina would react to these photons. However, our nervous system will filter out a single photon and will not pass it on to the central nervous system to deal with. Our eyes need to receive nine or more photons within 100 milliseconds. This meets the strength requirements for the photon-induced nervous system, and this signal will start the response of our consciousness: we can see these photons.

In other words, if we want to see the information of light irradiated into our eyes, a certain degree of intensity of the light is required.

In fact, from the point of view of anti-interference, if the eyes bring information of each photon for handling, then our eyes will continue to have the image of a mess, and our eye nerves will soon fatigue. Thus, minimum requirements for the input light intensity is not a weakness of the eye but a very clever system approach. This handling method is a noise filter.

Now it is clear, if we know the intensity of the light shining from Halley's Comet, according to formula (2), we can easily calculate the distance a pair of eyes must not be able to see Halley's Comet.

Of course, there are different eye visions. 1.11 eyes certainly see farther than 0.5. So whether the vision of the eye itself is good or bad is an important factor affecting the observed results. This is discussed in more detail later: the important concept of observable limited radius.

Limited observing radius of Halley's Comet relative to the eyes

Finite observing radius is an important concept we will define in this chapter and again use later.

Here, the observation of Halley's Comet with our eyes is used as an example to roughly explain the finite observing radius concept relative to the eyes.

If, with a pair of the best eyes, you can see Halley's Comet from within a radius at some distance but cannot see it a little further, this distance is called the limited observable radius of Halley's Comet relative to these pairs of eyes.

Similarly, with a telescope, the biggest distance the telescope can observe Halley's Comet is the limited observable radius relative to this telescope.

Telescopes can continue to be improved, and the corresponding limited radius can be continually increased. Thus, the limited observable radius we mentioned earlier, strictly speaking, should be named **Relative Limited Observable Radius (RLOR)**.

So, can the relative limited observable radius be endlessly increased?

No.

With improving observers from the eyes to the telescope to the space telescope ... the relative limited observable radius relative to one kind of object like Halley's Comet constantly increased, but this increase has a limitation. Eventually, no matter how much you improve the telescope, the relative limited radius will not become larger. This is the concept of **Absolute Limited Observable Radius (ALOR)** of a celestial object.

In the above discussion, the three objects involved in limited observable radius are the observation instruments (eyes or telescopes, etc.), the observing target (Halley's Comet, etc.), and the distance between these two.

For the Relative Limited Observable Radius (RLOR), changing the nature of these three objects affects the size of the observable radius, and for the Absolute Limited Observable Radius (ALOR), the telescope itself is not important; more important is the part that cannot be controlled by humans: noise. This part has not yet been discussed.

Noise Analysis

We know that during the observing process, there will be all kinds of interference noise.

Chapter 6 True Face of Hidden Celestial Objects and Dark Matter

The Chinese official translation of noise is noise sound, even in professional books that specifically describe astronomical telescopes, such as the very good Chinese book "Astronomy Visible Light Sensors" I am using. But within the scope of our current discussion, I felt awkward using noise sound. Instead, I have translated it as noisy wave.

There are various noises that act on the observation instrument. The sources of noises are both from the natural environment and from the observation instrument itself.

We divide these noises into two categories:

One is **controllable noises** that can be controlled by humans through various means, including such things as atmospheric disturbances, where we can move the observer up to mountains or to a satellite to overcome, or sunlight interference that can be blocked off, and so on.

The other is **uncontrollable noises**, of which there are two parts:

One comes from the nature of uncontrollable noises, such as diffuse light coming from the other celestial bodies when observing a planet, the background light of the sky, the zodiacal light, aurora light, etc. Statistical properties of the photon itself also makes the information received have ups and downs.

The second uncontrollable noise is non-natural noise. This noise is inherent to the observation instrument and represents insurmountable factors that optimization cannot overcome. One example would be the size of the telescope mirror: due to materials, processes, etc., it is difficult to make a large mirror with a one kilometer diameter, and even with a telescope that has a kilometer diameter mirror, we cannot reliably collect the weak information that is only one photon per ten square kilometers. And during the process of analysis, the received photons may be absorbed and reflected, or themselves cannot produce enough energy to record an event. Not all of the photons can be collected from the telescope to use; there must be some loss.

Here, we will generally divide noises into controllable and uncontrollable categories. That noise that humanity has no way to control is uncontrollable noise; that noise that we expect now or in the future to be optimized out to reduce its impact on the observations, therefore increasing the output of observations, is controllable noise.

If the optimization of human telescopes reaches uncontrollable noise level, then it's no use to continue to optimize the telescope. Specifically, for example, if we increase the sensitivity of a telescope to 10 photons per square centimeter, but uncontrollable noise is 20 photons per square centimeter, the receiving level of the telescope can only stay at 20 photons per square centimeter. It's useless to improve its reception level to 15 photons per square centimeter or 10 photons, i.e. it cannot exceed uncontrollable noise level.

Let's use our eyes as an example. If our eyes are good enough to see the level of four photons, but the noise level into our eyes is eight photons, then we do not see the four-photon information. We can only identify eight-photon information or more.

That is, since uncontrollable noise places restrictions on the observation of a light source, after the selection of the best human observation instrument, the maximum observable radius is decided by the uncontrollable noise. It is a fact that any human effort cannot change!

MATHEMATICAL ABSTRACT DEFINITION OF LIMITED OBSERVABLE RADIUS

Limited Observable Radius (LOR) of celestial objects is a very important concept and is divided into **Relative Limited Observable Radius (RLOR)** and **Absolutely Limited Observable Radius (ALOR)** of celestial objects relative to an observer. It can also simply be referred to as **Relative Observable Radius** and **Absolute Observable Radius**.

ALOR is decided by the luminous intensity of the celestial object that is being observed and the uncontrollable noise.

RLOR is decided by the luminous intensity of the celestial object, the controllable noise plus uncontrollable noise, and the observer optimization.

It can be seen that the 'Relative' of RLOR is named because there are factors that people contribute to.

We have already introduced a lot of concepts. Let's give them accurate mathematical definitions below:

The qualitative mathematical abstraction definition of Relatively Limited Observable Radius (RLOR) is this:

RLOR (C, NOISEnatural + NOISEcontrol, SENinst)

This abstract qualitative formula means that Relatively Limited Observable Radius (RLOR) is decided by the emission intensity LC of observable celestial object C, the uncontrollable noise NOISEnatural, controllable noise NOISEcontrol,

Chapter 6 True Face of Hidden Celestial Objects and Dark Matter

as well as observing instrument sensitivity SENinst together.

Through our research (see Appendix relevant papers), the more specific qualitative formula of RLOR is:

RLOR (LC, NOISEnatural + NOISEcontrol, SENinst)

$$= \sqrt{\frac{L_C}{4\pi f(\text{NOISE}natural + \text{NOISE}contral, \text{SEN}inst)}} \quad (3)$$

How to find a quantitative formula is the next research task, hopes on the reader.

When we consider the case of an extreme optimization, that is the controllable noise $\text{NOISE}_{control}$ and the instruments sensitivity SEN_{inst} in RLOR are reduced to a minimum level, zero, then the relative observable limited radius RLOR becomes absolute observable limited radius ALOR:

ALOR(C) = RLOR (L_C, $\text{NOISE}_{natural}$ + 0, 0)

Usually we write ALOR as LOR. That is:

$$LOR(C) = \sqrt{\frac{L_c}{4\pi f(\text{NOISE})}}$$

The concept of ALOR tells us that humans blindly trying to create bigger and better telescopes, after a certain time, has little meaning. As for the specific data, more in-depth study on the subject needs to be conducted.

This will make some people not like the topic. So we stop our discussing here.

Therefore, the fact that a telescope is bigger and more advanced does not necessarily mean that it's better. Good to a certain extent is enough, according to specific conditions.

DEFINITIONS OF HIDDEN CELESTIAL OBJECT (HCO)

Now, from the simple mathematical perspective, we determine the two most important formulas of the Hidden Celestial Objects.

formula. When the distance between the Earth and the observing celesti[al]
Earth) is greater than RLOR of the celestial object C, which is to meet

$$\text{RLOR} (L_C, \text{NOISE}_{natural} + \text{NOISE}_{control}, \text{SEN}_{inst}) < D (C, \text{Earth})$$

Then C is the hidden celestial object relative to the telescope.

As can be seen, the factors that make a Hidden Celestial Object (H[CO])
exactly the same as those factors that decide the Relative Limited Observabl[e]
(RLOR), namely the emission intensity L_C of celestial object C, uncontrolla[ble]
$\text{NOISE}_{natural}$, controllable noise $\text{NOISE}_{control}$, and the sensitivity SE[N of]
observation instruments, like telescopes.

Careful analysis of the changes in the parameters of the formula, that i[s]
sensitivity analysis for each independent variable, gives various classi[fications]
listed below, of the hidden objects.

No matter how advanced the telescope, for those objects that are o[utside]
absolute limited observation radius, those celestial objects are **Hidden** [Celestial]
Objects (HCO).

The celestial objects that are located outside the Relative Limited Ob[servable]
Radius (RLOR) but inside the Absolute Limited Observable Radius (RL[OR)]
called **Temporarily Hidden Celestial Objects (THCO).** Usually they a[re]
HCO.

From uncontrollable noise level, the Absolute Limited Observable Ra[dius can]
be calculated corresponding to a celestial object. We may never see a celesti[al]
located beyond the ALOR. The celestial objects located outside the [Absolute]
Limited Observable Radius (RLOR) are called **Constant Hidden Celestial** [Objects]
(CHCO).

Please note that whether a celestial object is THCO or not is rela[ted to a]
specific observing instrument. It may be a THCO for some telescopes bu[t can be]
seen by other telescopes.

The CHCO is relative to all observation instruments. If a celestial obj[ect can't]
be seen by the most advanced observer of the era, it is a Constant Hidden [Celestial]
Object (CHCO).

Trying to extend the RLOR by improving the quality of the observ[er is the]
theoretic basis for turning a THCO into an observable celestial object.

Chapter 6 True Face of Hidden Celestial Objects and Dark Matter

Summary of Hidden Celestial Objects (HCO)

Hidden Celestial Objects are invisible to us, but it is easy to prove their existence.

Around us, there are countless examples of real Hidden Celestial Objects.

There is a big difference in the quality of telescopes used to observe the sky. Those objects that can be seen with a high quality telescope lens, but which are not able to seen with a poor quality telescope, are Temporarily Hidden Celestial Objects to the poor quality telescope.

All comets, for people observing with the naked eye, or not using a super telescope, are periodically Temporarily Hidden Celestial Objects. The famous Halley's Comet is a specific type of Temporarily Hidden Celestial Object, for instance. Its periodic haunt writes a very clear footnote of Temporarily Hidden Celestial Objects.

But the simplest example is when we first watch the night sky with a telescope and then move our eyes away from the telescope and, with our naked eyes, see the same sky. And those celestial objects that were visible to our telescope but invisible to our eyes are Temporarily Hidden Celestial Objects to our eyes, and their existence is proved by the telescope used. It's that simple. The most notable example is to see the galaxies with our naked eyes. Our naked eyes can only see the Andromeda Galaxy (M31). This galaxy is 2.54 million light-years from Earth and is the farthest object we can see with the naked eye.

In other words, except the Andromeda Galaxy and the Milky Way we are in, all other galaxies are Temporarily Hidden Celestial Objects relative to the naked eyes.

Then extend our example to those galaxies that can't be detected by the best telescopes human beings currently have and are, thus, Hidden Celestial Objects. How many are there? How many of them are Temporarily Hidden Celestial Objects and thus will possibly be found in future because of the improvement of the observing systems?

At the farthest distance we can see, humans have only seen a small amount of particularly large celestial objects such as a Superstars. But in that place, according to the law of large structures of the universe, the celestial distribution there should be substantially distributed almost like our surrounding galaxy space. That is to say that the space there is also dotted with many star clusters, galaxies, stars, and planets. These are just normal objects that we cannot not see. These unseen objects are

193

Hidden Celestial Objects.

The above description defines many new concepts; here is a little summary, not only to clarify these concepts, but also to figure out how to define these concepts.

Did you not feel that the definitions of these concepts are completely natural and easy to derive? I was able to easily define them because of their own natural flowing out.

The thinking process is this:

1. Why can the naked eye see the distant Andromeda Galaxy but not see the nearby Proxima Centauri? Or why can we sometimes see Halley's Comet and sometimes not see it? ->

2. What factors influence the results of astronomical observations? What variations limit these factors? ->

3. Do our observations have limits? -> Limited Observable Radius

4. There are two extreme limit changes in different circumstances: the analysis of absolute factors and of non-extreme factors.

5. About changes of non-extreme factors, give a definition to describe it and name it as -> Relative Limited Observable Radius

6. About changes of absolute factors, give a definition to describe it and name it as -> Absolute Limited Observable Radius

7. What are the corresponding changes in non-extreme observation factors? Give a definition to describe it -> Temporarily Hidden Celestial Object

8. What are the corresponding changes in absolute observation factors? Give a definition to describe it -> Constant Hidden Celestial Object

In such a way, we defined our terms and obtained the following simple Figure 2.5, by which you can more clearly see the thought process:

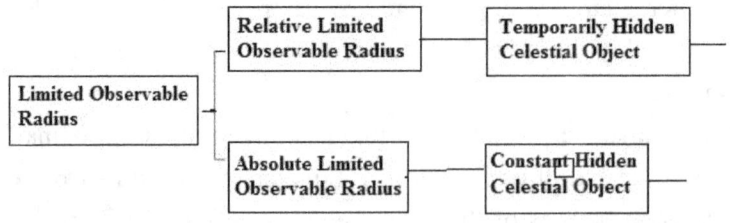

Figure 6.5 The HCO definitions extended from Limited Observable Radius

Chapter 6 True Face of Hidden Celestial Objects and Dark Matter

In the scientific research, I usually like, first and as far as possible, to submit the complex phenomenon to the processes of abstraction and generalization. First, we do analysis of changes in the most general cases; then we do analysis of those possible changes in the extreme cases. I hope this simple scientific method can give you a little inspiration in your thinking.

The general idea of scientific research

I am going to deviate from astronomical themes for a moment and do a little nagging. The purpose is to share a little of my research experience with you readers - especially young friends.

Above is a general idea of doing research that helps us be able to abstract what we need from complicated reality that is perhaps important to scientific results. This is making abstract and summarizing, general scientific method.

Another point is more important: the discovery of new ideas from special information received.

Here is one of my own experiences as an example, at the time I was working on my master's thesis.

My specialty is optimization design and bias computer simulations. During an internship for graduate student research, I had actually been to only two coal mining work sections, but I optimized and rewrote the best section length of the textbook for comprehensive mechanized coal mining. That paper was published in the Coal Science and Technology magazine as the "coal mining experience summary."

The idea that made the rewrite the conclusion of the coal mining textbook, is for my master's thesis, came from a newspaper article. The newspaper reported that the mechanized mining team of Datun Coal Mine in the 96 meter-long working, set a record high for national mechanized mining production.

At once, I saw that there was something not usual: the textbook on coal mining clearly wrote that the optimal length of fully mechanized coal mining section should be about 180 meters, not less than 120 meters. Otherwise, with the reduction of the section length, the productivity will be decreased. That is, the mechanized coal mining section should have a sufficient length to allow coal mining machinery to work efficiently. And subject to geological conditions, many mines cannot be arranged in a long work section. Therefore, if the short working section can achieve high efficiency, it is a major significance to the coal mines whose geological conditions are not good.

Now, in Datun Coal Mine, the length of its mechanical mining work section was limited by the geological conditions, only about half the optimized length that the textbook defined. Therefore, its work efficiency should be very low. How was it possible for them to achieve the nation's highest productivity record?

So I went from Beijing to the coal mine to investigate why and how, and also to collect some data for my computer simulations project.

It was found that they were using a central infeed mode, which is good for those coal mines whose geological conditions are not good enough to lay out long coal mine sections.

I put it into my master's thesis as one chapter.

Interestingly, at the time of my master's thesis defense, the director of the coal mining lab who attended my defense raised serious doubts and controversy, but later, after the paper published in Coal Science and Technology magazine, he found me and apologized. Now, these years, do such kind events as a professor apologizing to the student ever possibly happen? Luckily, I had a good professor. Also, we see how hard it is to modify the conclusion of a textbook.

There are two reasons I meant to write this little story. One is to instill the idea that, when engaged in scientific research, it's important to have a sensitivity for special information that is seemingly unrelated to what we are working on, unusual information, and to use our brains for deep analytic thinking based on basic scientific truths. Do not rush to question or deny and don't be indifferent; otherwise, you may miss the opportunity for new discoveries. The second has to do with the authority that is to be broken; how can we do good research without breaking authority?

This, of course, also applies to the usual "bold speculation" approach to cosmology. But we have to be more carefully "bold" and think carefully with detailed confirmation.

Young friends, hope is in you! I hope my book will sow the seeds of an authority defying tree so that you can grasp a little more correctly the scientific method of increasing a little inspiration. Trees grow into the top, days in the future.

Hidden Celestial Objects are not dark matter

According to NASA, a new study (November 24, 2015) reported that the Earth might be surrounded by dark matter "hairs": the most intensive roots. Surrounding the earth, may be dense dark matter filaments. Dark matter is a mysterious, invisible

Chapter 6 True Face of Hidden Celestial Objects and Dark Matter

substance, making up approximately 27% of the universe of matter and energy. This is a new progress in human understanding of dark matter.

We know that the definition of dark matter is matter that does not emit any light and does not reflect any light, but has gravitational field. Is there such a magical substance? What kind of molecular and atomic structure would such substance has, that makes it be indifferent to the outside world, there is no the slightest communication channels with the outside world except gravity? As far as our experience the people have on the Earth, it seems there is no such thing with such a self-imposed exile. This "dark matter," except for the Earth "hair," so far has not been identified any entity and still exists only in the minds of cosmic physicists.

The discovery of Earth "hairs" may cause the change of the definition of dark matter, right? Otherwise, what gravity can the "hairs" have? I do not know how invisible "hair" of matter can be shown by the photo. I hope NASA experts can make such a theory completely rounded.

Figure 6.6 NASA reported that the Earth is surrounded by "hair-like" dark matter

Cosmologists also have claimed that the universe is filled with dark matter and dark energy, that 96 percent of the universe matter are them..

But if celestial dark matter or dark energy is just these "hairs," the quantity of the dark matter or dark energy is too little. And people may only see this suspected dark matter "hairs" around the planet and may not see it around the light-emitting stars. Then, if you total up the amount of dark matter in the universe how much would be this substance?

In the past, the only way finding this dark matter was to use its gravitational field.

Now, how can we find the "hair-like" dark matter? By taking photos? Can we take pictures of dark matter? Currently, we haven't seen the news reports from the experts, so we stop doubting about it now.

The use of the gravitational field to find dark matter is nothing more than making direct measurements of the gravitational field of dark matter. This seems to have made little progress so far. The other way is like the discovery of Sirius B: by measuring the influence on the orbits of Sirius. In the same way, dark matter can be found by observing its impact on nearby luminous celestial objects and then extrapolating its existing. But dark matter objects found in this case, until now, were all finally proved to be ordinary Hidden Celestial Objects. No exceptions.

No exceptions! I don't understand why the cosmologists still insist that there is this mysterious dark matter but that there are not normal celestial objects that can't be seen directly.

In recent years, the astronomical community has spent a tremendous effort to find dark matter. But, in fact, all objects found so far that were once thought to be dark matter objects turned out to be mistakenly marked Hidden Celestial Objects. After finding such a case, it was rarely mentioned that it was wrongly labeled as "dark matter" but ordinary celestial object.

The most famous example was the discovery of the companion star of Sirius, Sirius B. Figure 6.7 is a screen shot of the relevant reports. This is also an example of finding a hidden planet by examining its gravitational influence on the orbit of other objects. It found the first well-known instance of "dark matter" (actually, it is a hidden celestial object).

In Fig.6.7, in his paper titled "Existence and Nature of **Dark Matter** in the Universe," Virginia Trimble, who was working at the University of California, Irvine, and the University of Maryland, reviewed the 1987 discovery of Sirius B: "The first detection of non-luminous matter from its gravitational effects occurred in 1844, when Friedrich Wilhelm Bessel announced that several decades of positional

Chapter 6 True Face of Hidden Celestial Objects and Dark Matter

measurements of Sirius and Procyon implied that each was in orbit with an invisible companion of mass comparable to its own." Because, at that time, the telescope was relatively backward, the telescope could not see Sirius B, so it was defined as dark matter in the report title. But with the improvement of the telescope, people were later able to see Sirius B through a telescope. Of course, after that it is no more the mysterious dark matter.

Starting with Sirius B, all so-called dark matter that people have found through its effect on the gravitational field have now been identified as temporarily Hidden Celestial Objects.

The astronomical community generally believes that 96 percent of the universe is dark matter. If this assertion is correct, then a huge majority of dark matter can only be hidden planets or galaxy because everything that was once thought to be dark matter was later confirmed as Class 4 temporarily Hidden Celestial Objects and not dark matter. With the change of the observation instruments they went from being hidden to be exposed to our vision. Consequently, we could make the following hypothesis: most, or even all, of dark matter actually are Hidden Celestial Objects!

EXISTENCE AND NATURE OF DARK MATTER IN THE UNIVERSE

Virginia Trimble

Astronomy Program, University of Maryland, College Park, Maryland 20742, and Department of Physics, University of California, Irvine, California 92717

1. HISTORICAL INTRODUCTION AND THE SCOPE OF THE PROBLEM

The first detection of nonluminous matter from its gravitational effects occurred in 1844, when Friedrich Wilhelm Bessel announced that several decades of positional measurements of Sirius and Procyon implied that each was in orbit with an invisible companion of mass comparable to its own. The companions ceased to be invisible in 1862, when Alvan G. Clark turned his newly-ground $18\frac{1}{2}''$ objective toward Sirius and resolved the 10^{-4} of the photons from the system emitted by the white dwarf Sirius B. Studies of astrometric and single-line spectroscopic binaries are the modern descendants of Bessel's work.

Figure 6.7 the story of finding the companion star of Sirius, Sirius B, and the first case of "dark matter" reported. Of course, this kind of "dark matter" is false. The title of this figure gave the definition of the dark matter.

There is additional theoretical basis to support such an inference: the magical logic used by NASA to draw the microwave background whole sky map, which is the law of the universe or the law of large structures. Later, we will carry out a detailed discussion of the whole sky map.

THE HIDDEN CELESTIAL OBJECT CONCEPT IS MORE PERSUASIVE THAN THE CONCEPT OF DARK MATTER

Hidden Celestial Objects are real. We can see them once their hidden faces are exposed through a variety of historical events. We can see their naughty looks expose themselves and then disappear from our sight, and can also see their cycles on the catwalk.

Dark matter, so far, exists only in reasoning and imagination. And, by definition, dark matter is a strange foreign substance that there is no way to detect but only by gravitational effects. Dark matter is the kind of substance in need of people to have a "special understanding"; it's not so strongly persuasive.

It has been asserted that more than 96 percent of matter in the universe is dark matter. In the discussion section below, we can see through the application of NASA's data that the vast majority of matter in the universe is made up of Hidden Celestial Objects instead of dark matter. We can say that **more than 96 percent of matter in the universe are Hidden Celestial Objects. So, where is the 96 percent of dark matter and dark energy?**

All these are indicating to us that the vast majority of dark matter is actually normal, ordinary, reflecting or glowing Hidden Celestial Objects. The existence of dark matter is not only strange but suspicious and, so far, has not been confirmed. The fact that there are no more specific examples except "hairs" is also not convincing. Therefore, even if such a strange matter exists in the universe, it is only a little handful.

Hidden Celestial Objects categorized

This categorization follows the classification concepts in Figure 6.5, sensitive analyzing the relationship between all relevant factors of Hidden Celestial Objects in its definition formula and considering increases or decreases in these factors caused

Chapter 6 True Face of Hidden Celestial Objects and Dark Matter

by change, resulting in the following Figure 6.8.

For more detail about specific examples and the relevant detailed analysis on various kinds of Hidden Celestial Objects, please refer to the appended paper.

I'm here to specifically talk about the Hidden Celestial Objects of Class 10 at the lower right corner in Figure 6.8, whose light intensity is a constant 0.

This situation is depicted in the class of phenomena known as "dark matter." The official definition of dark matter is an object that does not emit any light. Zero light intensity of dark matter is in its definition, and the relative observable distance is zero; thus, dark matter, to any observer at any distance, is a hidden object.

Some similar planets also belong to non-luminous celestial objects, like the moon, the earth, and the like. But they absorb light, reflect light, and thus will glow with a weak light. When some of them are within an appropriate distance to the Earth, they may be observed, becoming temporarily Hidden Celestial Objects. When they are outside a certain distance, because the reflected light is weak, it is difficult to observe them. Thus, with the improvement of telescopes, or with the use of a space telescope to probe the universe and go deeper in our observations, then even if we find new planets all around us, there is no need to fuss.

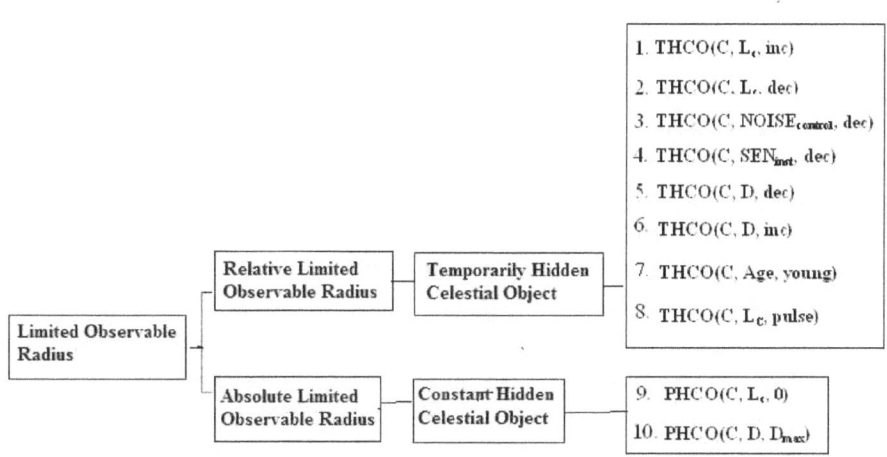

Figure 6.8 illustrates the theoretical system of Hidden Celestial Objects

And this is the foundation of the idea that is possible to find alien bases or find planets that we are able to immigrate to.

The imaginary cosmic picture of the new concept

We use celestial data provided by NASA, combined with hidden celestial theory, to construct a new celestial distribution of space dimensions schematic Figure 6.9.

Figure 6.9 uses the celestial data provided by NASA. IC1101 is the largest celestial bodies discovered so far, we can calculate its limited observable radius is 477 billion light-years, which means that basically IC1101 light may spread to 477 billion light years. We use it as a theoretical maximum distance that can be observed by human in Figure 6.9.

So you can see, in the figure D_{NASA} is the size of 13.7 billion light years that was determined by NASA in the Big Bang model, while IC1101 light can reach is 477 billion light years. That is, according to NASA observation data to infer, at least from 13.7 billion light-years away of Earth, to 477 billion light years, in the huge space all the celestial objects are belong to hidden celestial objects relative to human on the Earth.

Haven't seen article that explains how the Big Bang barrier may stop IC1101 light cross this barrier, and what would happen at that time. I do not know now what the relevant thinking from the universe astronomy academic.

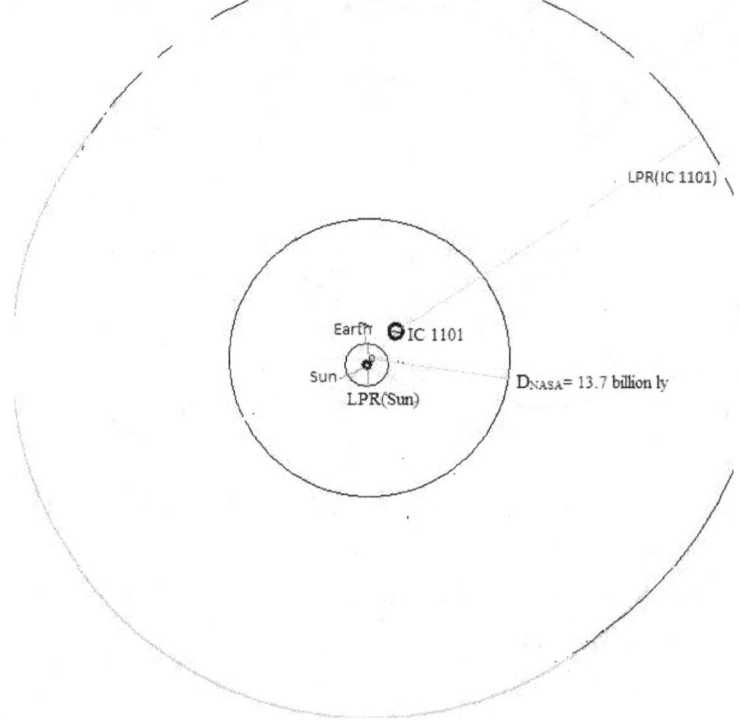

Chapter 6 True Face of Hidden Celestial Objects and Dark Matter

Figure 6.9 The imaginary cosmic picture of the new concept

Also, according to the theory of cosmic large scale, the identified areas on the edge of the Big Bang should also have many similar Milky Way galaxies, sun-like stars and other celestial objects, do we know when all of these celestial light shine on the edge of the Big Bang barrier, what will happen?

In contrast, the size of the universe according to the Big Bang theory of the experts given, the size of 13.7 billion light years is too insignificant, and that's even less than 0.03 of the distance the IC1101 light may shine to.

Now we can roughly describe what the image of the cosmic panorama summary in our mind

The universe is huge, immense, and we never know where its borders are. Before we can travel at the speed of light and extend the life to Hooray for the unit, any attempt completely uncover the mystery of the universe is in vain. Perhaps only the gods can!

The universe we can understand is currently limited to a relatively small radius of the range, which is within NASA official certification of 13.7 billion light years away, and it also cannot be complete determined, but basically to say, human understanding of the universe is reaching the limit. Please note that not like Hawking said, is that the human almost know everything of the universe, the human understanding of the universe was almost reached the limit; on the contrary, the ability of human understanding of the universe has basically reached the limit, under which limit the ability of humans can only know a little section of universe.

Human is limited his steps within the range of less than a light-year; the human eye can't see through the thick veil of distance.

The universe that we can maximize our perception is based on the relative limited observable radius of the maximum luminous celestial object. Currently it is defined by the IC1101 data shown in Figure 6.9. This can be changed, but until today, probably will not have much of a breakthrough. And to calculate with IC1101 data, it is also a very small size and ratio comparing the universe size of human claims to already known observable radius of IC1101. Humans are confined to a relatively insignificant area of the universe. The proportion of the universe known as the human understanding and the true size of the universe, is a less than .000 ... % proportion. And those unseen celestial objects are hidden celestial (we cannot see but there is a celestial body), so there is no room for the 'dark matter' presents in a large proportion!

The distribution of celestial objects in the universe may be follow the pattern we now understand, according to different distances between the celestial, but we're not sure of it. We do not know beyond the range of human perceptible universe will transform its existence with different form and appearance.

Concept distribution picture of Hidden Celestial Objects in the universe

We use above imaginary cosmic picture of the new concept to construct a hidden celestial objects distribution schematic diagram 6.10.

NASA released the observed data of the universe, all-sky map, etc., are shown that NASA took the Earth at the center of the universe. However, according to NASA data, the Earth is more than four billion years old, while the age of the universe is 13.7 billion years, even follow the Big Bang theory, the Earth cannot be placed in the center of the universe.

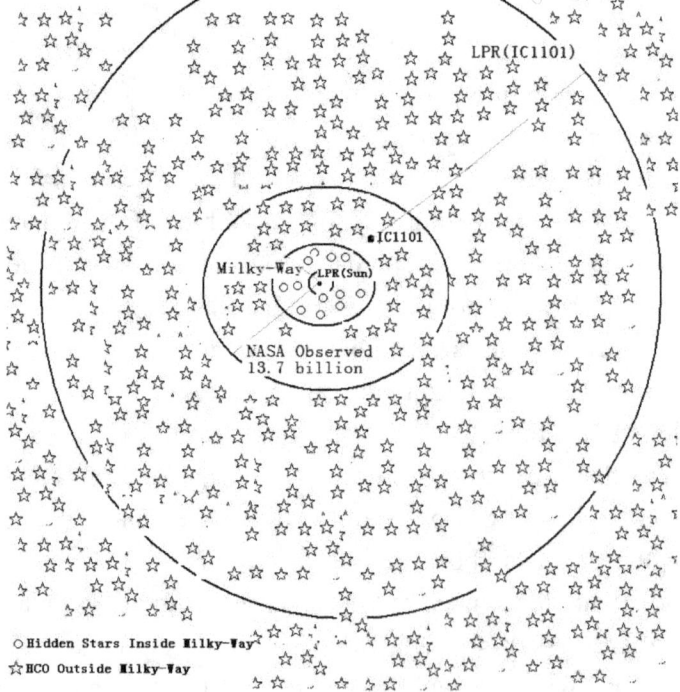

Figure 6.10 hidden celestial objects distribution schematic diagram

But from another point of view, from the theory of limited observable radius of celestial objects, the Earth is in the center of the observable universe. Because we use telescopes to explore the surrounding space, we can only see a limited observer centered universe! And this is from the other side to prove the correctness of the hidden celestial theoretical system.

The hidden celestial concept makes the existence of nearby alien bases and immigrate planets possible

The distribution of hidden objects can be seen from Figure 6.10, which shows that a lot of hidden objects and planets are also distributed around the earth. In these hidden planets, there may be a great fit for human habitation.

These hidden, near-Earth objects can be discovered by intensively searching with advanced telescopes. Because the area to be searched is relatively small, in the range of several light-years around the earth, and telescopes are powerful enough to be used, there is hope that we can find Hidden Celestial Objects that we previously have not noticed or did not see. These newly discovered objects are planets that we can likely immigrate to, or are possible alien bases. If aliens and earthlings are more or less the same in essence, then the planet they live on would require the same habitable conditions that humans need and that are present on Earth. In other words, if we find an alien base, we might actually also find a planet that humans can move to.

Hidden celestial bodies are a possible source of the cosmic microwave background

It is possible that hidden celestial bodies are a real source of the cosmic microwave background. Maybe just a few objects, maybe several groups of objects, but there is this possibility.

Because the Earth is in the vicinity of the microwave unknown sources that have been detected and because, at this point, no one has been able to find the source of the microwave radiation, it has been put down to being the remnants of the Big Bang. The fact that satellites around the earth probe for a few years before saying that microwave radiation is almost the same everywhere in a universe billions of light-years across, while also adding the term "background," speaks very little of the experts' use of basic scientific principles.

This topic is left to be described in detail in the discussion of the whole sky map of the cosmic microwave background.

Conclusion

Dark matter is one of the hottest topics in cosmological theory and the one that most needs to be addressed.

Dark matter is one of the most peculiar substances in the universe. By known atomic and molecular theory, we do not know how to construct a non-luminous, non-reflecting, magical material that can also absorb light.

In scientific inquiry, for "strange" things, we need to be skeptical. For something in violation of the basic principles of science, we want to treat it with caution.

The alien base topic was only an introduction that extended the concept of hidden objects. However, the hidden celestial concept can solve the problem of the existence of alien bases near Earth and also truly offers new, feasible ideas for migrating to alien planets.

We now have clear evidence that celestial objects lie hidden from our sight.

Our eye comet idea is composed of and supported by basic scientific theorems. There is no mystery, no absurd argument for "mysterious science."

Solidly validating our ideas with basic scientific principles, experimental science, and analytic reasoning, we bring with our smart eyes a truer, more solid face to the universe, and out of our reach vaguely shadow of its back。

CHAPTER 7

FREQUENCY RHYTHM OF TINT AND RIPPLE

- The real secret of the speed of Observed Image and the Hubble's Law[5]

The nature of sound and color change is the change in the wave frequency.

As wave frequency gets higher, the tone gets higher, and the hue gradually changes blue; as wave frequency gradually gets lower, the tone gets lower, and the hue gradually turns red.

Change in wave frequency is independent of the strength of the wave.

If we do not think deeply, then our understanding will stop there. In explaining certain physical phenomena, we may therefore go astray; what we see is the appearance rather than the essence of the problem.

Under certain conditions, changes in the strength of the wave causes the wave frequency change unexpectedly, playing a wonderful role.

In exploring the causes of celestial redshift, we note the following: the nature of the wave is decided, and the change of the power of the wave will not cause the change of the frequency of the wave.

We know the Hubble Constant is closely related to the Doppler redshift. Thus, we cannot fail to note the strange Hubble Constant values obtained in astronomical observations. That is, when receiving the same celestial light on the earth at the same time, just because of the different observation instruments, the values differ by as much as two to ten times. This is completely contrary to the nature of the wave. Different observation instruments can only lead to different strength of the received light waves but cannot change the frequency of light!

---

[5] Refer to appendix "**Speed of Observed Image and the Observer Caused Redshift of a Celestial Object**"and the references of the paper

In most cases, this would have been at loggerheads over many years, and eventually someone would need to think of another point of view or theory to explain this phenomenon. But NASA gave strong support to its telescope and with one invincible attitude and suppressed all the different observations. The victor told the astronomical community: It's my best telescope, so it's my right! So all the different debating voices are suppressed, which, of course, killed the possibility that people in dispute could eventually go to find what the real reason might be. With advanced instruments, NASA wiped out the possibility of discovering the truth.

But advanced instruments have no mind and no thought. Differences and anomalies remain. Out of the heap with taxpayers' money, NASA's astronomical telescope cannot eliminate or explain them.

When we notice that different observation instruments lead to huge differences in frequency, we will naturally go to explore this phenomenon, and thus include the performance of the observing instruments into observational results analysis theory rather than deal with a variety of topics such as calibration, repair theory, and so on, which would only be correcting the wrong theory.

From here, we will find the scientific theory that allows a variety of different observation results to be analyzed and summarized, with no need to argue which are correct and which are wrong!

We will master the golden key for us to open the true mystery door of Hubble's law.

Alternate between tint and ripple

Close our eyes, sound comes into the ear; open our eyes, sultry color. The two phases of sound and color are born with people. Slightly erotic and sensually suffused, the "king of entertainment is not a sensual, useless, hard-to-learn waist dance." Internally awake, tint and ripple are vivid, "will make the pleasant aqueous delighted mountain, because obtaining the true vivid dance and song."

Sound and color, although seemingly very different, in reality are as one pair of twins, brother and sister, with many similarities: they are waves. Sensual changes are caused because the frequency conversion. With just a single frequency, sound would be without accent or grace, and color would lose its brilliance. Therefore, the elegant rhythm of frequency variations provides gorgeous, sensual changes.

The number of features shared between them makes them see naturally related;

if one masters some relevant acoustic wave characteristics, he can transfer this mastery to the use of natural light wave.

This is a seemingly very natural conversion, so we shouldn't have to undergo a rigorous proof of it.

But the problem is precisely in this "natural" connection!

It's such a simple formula, so easy to understand, that we feel like we do not have to think carefully about it, and sometimes, people slip into the trap without hesitation because of its simplicity, like the example discussed later about Einstein applying relative concepts unthinkingly to elementary mathematical formula; with "natural" reasoning, whipped in from here to there, people often smoothly and consciously go astray. Another example is how the understanding of how changes in acoustic wave frequency can be due to the wave's characteristics was extended to natural light without considering the exception phenomena.

So tint and ripple not only let monks tremble vigilantly; color and sound also can let scientific research lose its luster.

FROM SOUND WAVE TO LIGHT WAVE-- FROM THE DOPPLER EFFECT TO HUBBLE'S LAW

This chapter discusses two major problems:

The first is the actually used time of the light from distant objects after a long journey, which is transformed into useful information before finally being transmitted to the viewer. This is an important issue that the astronomical community ignored. This problem is solved by issues such as the method of telescope information handling, celestial light intensity with distance propagation attenuation law, dealing with the impact of time used for the telescope information handling, and all of these factors, to produce a comprehensive analysis and conclusion of their variations.

Then use the above results obtained to re-interpret the celestial redshift phenomena and laws.

First, a quick look at the relevant knowledge.

From an article on "Hubble's Law" in the basic "Baidu Encyclopedia":

All the celestial bodies in the universe are in motion; the celestial astronomy space velocity on the observer's gaze direction is called radial velocity component objects. Based radial velocity measured by Doppler Effect in physics, Austrian physicist Doppler (J.C. Doppler) first discovered it in 1842. The effects noted, the movement sound emitted by the sound source (such as high-speed movement of the train whistle), the stationary observer listening to changes. In terms of c represents the speed of sound, v is the velocity of the sound source, the stationary observer actually heard movement in the wavelength of the sound source audible λ, when the sound source stationary relationship between sound wavelength λ_0 comply mathematical expression ($\lambda-\lambda_0$) /λ_0 = v / c, known as the Doppler Effect. Because the speed of sound c and still wavelength λ_0 are known, λ may be determined by actual measurement, it is possible to use the Doppler Effect to measure the speed of the sound source movement v. The higher the velocity of the sound source, the acoustic wavelength change is more significant.

Light is an electromagnetic wave, so if the Doppler Effect is equally applicable to the dissemination of celestial light, the formula c is the speed of light and v is the radial velocity of celestial bodies. Stars, for example, usually have some stellar absorption lines in the spectrum, which is emitted from the surface of the starlight radiation is absorbed various elements in the star's atmosphere caused by strict and specific elements corresponding to a specific wavelength absorbed several pieces line. As long as the absorption lines in stellar spectra measured position of an element (i.e., the movement of the light source wavelength λ), and the laboratory standard line elements in the same position (i.e. stationary wavelength λ_0) comparison, you can find both inter will have some displacement $\Delta\lambda = \lambda-\lambda_0$, i.e. Doppler shift. λ_0 is known, and $\Delta\lambda$ can obtained by observing, so by the Doppler Effect to calculate the star's radial velocity v, which is basic principle to determine the radial velocity of a celestial.

It is seen from the above description that, because "Light is an electromagnetic wave," the astronomical community naturally extended "The Doppler Effect to the spread of celestial light" with little thought or doubt.

This extension has not been proven and is a possible source of the error.

IF THERE IS A DIFFERENCE BETWEEN SOUND AND COLOR, WHY DO WE NEED A REINTERPRETATION OF CELESTIAL REDSHIFT?

Celestial redshift (hereafter abbreviated in this book as redshift) is the phenomenon of astronomical telescope spectral lines shifting toward the red end.

It has been recognized as redshift caused by the Doppler Effect.

Although there are many people looking for a non-Doppler redshift mechanism to explain the effect, no one has noticed yet that the redshift values received at the same location on different instruments from same celestial object were different from each other. With the different observation instruments, the obtained redshift values changed; the higher the sensitivity of the instrument (such as the Hubble telescope) the smaller the observed redshift values.

The value of the Hubble Constant was once a long and intense debate topic. After half of the twentieth century, the value of the Hubble Constant H_0 was estimated between about 50 to 90 (km/s)/Mpc. Géǀrard de Vaucouleurs claimed it should be 80 and Allan Sandage believed it should be 40.

In 1996, a debate between Gustav Tammann and Sidney van den Bergh was held, presided over by John Bahcall, the theme of which was the two competing values of the Hubble Constant.

In 2003, using the results of the highest precision WMAP cosmic microwave background radiation, the measured value was 71.4 (km/s)/Mpc, and then in 2006, 70.4 +1.3/-1.4 (km/s)/Mpc was the measured value.

Historical values of the Hubble Constant show that the 1920s measured values were up to ten times greater than those measured in the 1980s (Figure 7.1). Until 1996, there were intense debates on the Hubble Constant being 80 or 40, which in any case should not appear in the Doppler redshift, and neither, obviously, should simply be explained by instrument error or calculation error.

We believe that the scientists who took the observations, no matter who they were, without doubt were of good scientific quality. In the long term, the collective was divided into two schools of debate, and everyone on each side of the debate must have done good preparation and carried out the most serious scientific observations. But because of the different observation instruments used, there was more than a double phase difference in their results, which is an incredible thing.

Let's change the direction of our thinking.

What if everybody's observations were right? Is there such possibility?

If everyone's observations were right, then does it mean that the theory applied to explain the results was wrong?

In this respect, if people had never thought about it, it is now time for people to think about.

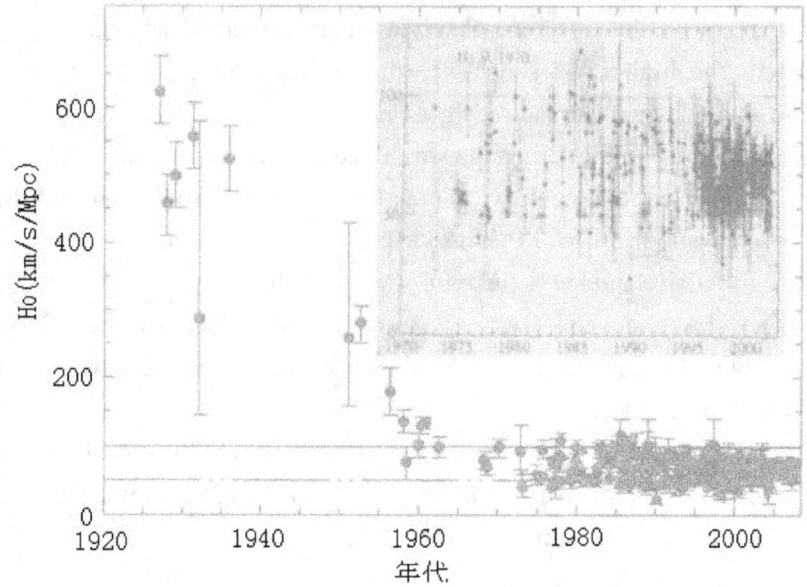

Figure 7.1 different values of Hubble Constant H0 measured in different years

On the one hand, according to Hubble's law, the wave motion will change the frequency, resulting in redshift of light from celestial objects. On the other hand, the light waves reaching the Earth are the same to any telescope used for receive them. So the optical frequency change is caused by the good or bad characteristics of the receiving instrument itself, which is not consistent with the scientific principles we had.

The simplest example: in the opera theater, put your hands over ears while you listen to a faint song. Will the song out of tune? At the same location, listen to the same song with very good and very bad receiving antenna band radios. There will be a big difference in the volume of the song between the two, but it does not appear to make the song horrendous. A difference of more than ten times the quality may not be present in two different receivers tuned to the song. With the worst and the best receivers in the same place receiving the passing train whistle, the volumes received will be significantly different, but the siren sound frequencies received by the instruments cannot be significantly different.

The Hubble Constant just reflects the celestial light frequency changes. Thus, according to the above-mentioned acoustic characteristics, it makes us start thinking: if cosmological redshift is not caused by the universal expansion movement, then what is the real reason that led to astronomical observation of redshift?

Historically, there have been scientists with different views on the Doppler redshift. The most famous one was the "tired light" theory put forward by Fritz Zwicky in 1929. Zwicky believed that photons would slowly lose their energy in the huge transmission process over great distances in steady state space due to continuing interaction with each other or with other substances, or due to the action of some unknown physical mechanism. Because photon loss of energy corresponds to the increase in the wavelength of light, this effect will make the redshift increase with increasing photon travel distance.

But this idea and others like it have not been recognized by the mainstream scientific community. They are referred to as non-standard cosmology.

So why do we now want to put forward a new interpretation of the redshift mechanism?

Mainly the following reasons that no one has ever introduced:

- There should not be a huge difference in Doppler redshift values between different instruments of different sensitivity. Sometimes while listening to the distant sound of singing, even though it is intermittently faint, the part we hear is still substantially the same pitch. Tone does not have to multiply the difference. Over time, the Hubble Constant has seen a huge difference in values: depending on the different instruments or methods of observation, the frequency of the received light has changed dramatically. In particular, in the historical debate held on the big differences of numerical Hubble Constant values, the two factions fought over whether the Hubble Constant is 80 or 40, double the difference. This cannot help but make you wonder whether the redshift is actually caused by the Doppler Effect. And these differences are explained satisfactorily by our theory of observed image speed changes.
- So many scientists, for such a long time, were confident that their observations were correct. Then who was wrong? Everyone was right! The theory behind the interpretation of these observations was wrong! Correct theory should be able to explain all of these different observation results obtained from different telescopes!
- All redshift values are generated by the observation instrument. As can be seen from Figure 7.1, the poorer the quality of the telescope, the greater the value of the Hubble Constant obtained, which says that the Hubble Constant has a certain relationship with the quality of the telescope and varies with different instrument sensitivity. It is a function of the sensitivity of the instrument. So far, no theory in the analysis and calculation of the redshift factors takes instrument quality into

account. But what data can be obtained without going through an astronomical observation instrument (telescope, or the human eye, etc.)? At least they should be taken into account as one of several factors affecting the Hubble Constant. Even if this difference was simply a function of instrument noise, the observations of redshift data should still not deviate so hugely based on the different instrument used. All of these theories, in turn, can use our observed image speed changes to explain the difference.

- Bond et al. study in March 2013, said the age of the universe is about 13 billion years and that HD140283 is only 190 light-years from the Sun. If you follow the Hubble law, in a constantly expanding universe, how can HD140283 stay only 190 light-years from the Sun?
- According to Hubble's theory, very close objects do not have redshift. But virtually all the stars, including the Sun, have redshift. Hubble's theory cannot explain this phenomenon, so the gravitational redshift theory was used to fix the problem. But our observed image speed change theory can satisfactorily explain all phenomena.
- Using filters will lead to changes in the value of the redshift phenomenon. This is actually because the light entering the telescope is weakened by filters, thus causing a change of redshift values after the light intensity decreases. This phenomenon is completely independent of the frequency change of the light, but is only related with the change of the light intensity. Hubble theory cannot explain the phenomenon, but observed image theory can.
- One test comparing the historical observation data and modern data would be to check if there were any celestial objects that were identified in the past that have now disappeared. Human celestial telescopes have recorded more than 300 years of observations. Using the same telescope from years ago, under the same conditions, observe and compare to see if there has been a large number of now missing celestial objects. If this situation does not occur, it indicates that expansion of the universe theory is not correct because, under conditions of expansion of the universe, according to Hubble's law, distant objects move away from the Earth at a very high speed. Several hundred years after the movement, some celestial objects must have become hidden. This experiment is not difficult to do; it just requires manpower and time.

All of the above prompted us to put forward a new theory on the cause of celestial redshift - based on the new theory of celestial observed images speed, the main reason to reinterpret celestial redshift.

We used the new theoretical point of a new theory to explain the celestial redshift, and the results obtained were better because we explained and resolved all contradictions in the original theory, specifically resolving the big different values obtained from different observation instruments. Our theory can reasonably explain why those big differences come from different observation. Instead of only using the data from NASA's most advanced satellite telescope to calibrate the different observation results, all different results from different observers were treated as valuable.

In classical physics method, we systematically studied the effects of the composition of the following three things: the observed celestial object, the starlight propagation character and distance, and observation instrument sensitivity and response time or integration time. From these, we determined the system effect of the observed image speed change is the main mechanism of celestial light redshift.

Finally, we designed an experiment to test our new theory. By simple observation, we can determine which theory is more correct. We hope astronomical observatories would like to do this test. This is likely a test to lead to a major breakthrough in the theory.

THE SPEED OF LIGHT, THE SPEED OF CELESTIAL LIGHT, AND THE SPEED OF CELESTIAL IMAGE

Some definitions about the speed of light, which Einstein defined as the maximum speed of the universe, will not be discussed here.

So, how about the speed of celestial light?

Usually people measure the speed of light in a laboratory on Earth, and confusedly mixed the light speed measured from a lab with the speed of starlight.

But in fact, on closer examination, the speed of celestial light and the speed of light we are now commonly using are very different.

In 2006, I published an article entitled "The Speed of Photons and the Observed Light Speed from a Star", which was the first article that mentioned that overall statistical light speed emitted from the light-emitting celestial is not the same as the speed of single photon.

Einstein said that the single photon velocity of light can be accurately measured from a laboratory, about 300,000 kilometers per second or so. This is what people commonly use for the speed of light and is the only value recognized by the scientific community.

After long-distance transportation, the starlight arrives in the vicinity of the relative observable limited radius of the telescope (see the "Hidden Celestial Objects" chapter - the relative observable limited radius was defined there - and the "different functional regions of photon batches' section below.) Since the same number of photons emitted by a celestial light wave at that distance is insufficient to maintain the surface on which these photons formed, it is necessary to send following batches of photons to supplement, which therefore makes the star's speed of light propagation overall relatively slower near the relative observable limited radius. That is, the speed of light from far away stars becomes slower than the laboratory measured light speed!

Is it likely to be a problem with Einstein's speed of light theory? After the above-mentioned article was published, I was still thinking about this problem, always feeling a little less critical of the middle of things. In the past decade, thanks to Johnny and his discussions about the above idea, there have been some new breakthroughs.

In the summer of 2014, we took Johnny, who was just admitted to the University of California, to Northern and Central Europe for two weeks to celebrate the end of his high-school life, that is, the end of his rebellious youth that gave us headaches, before he briskly headed off to the larger world.

One day while shopping around in Amsterdam, Netherlands, I spoke to him about the point of view from my paper "Photon Speed and the Speed of the Stars Observed" written 8 years ago. After listening, Johnny immediately replied: "It's not the star photon's speed that has changed, but the observed image that has changed speed. Einstein is not wrong, but the calculation of the speed of the observed image is!"

His words caused me some thought, so I was able to jump out of the trap of established thinking and into a new perspective to re-examine and think about this problem.

We were discussing this issue on a busy street of Amsterdam. In Amsterdam, exotic scenery quickly becomes casually superficial.

Johnny and I often discuss some of the issues I consider. The main idea is to take away his time and put the excess part of his adolescent energy toward research. He had worked with me on the subjects in this book later for two consecutive summer vacations and helped me to deepen the ideas and write the research papers. Many of his views amazed me. The Hidden Celestial Object concept was completed with his help. He wrote the English papers.

I also discussed with him my views on the theory of Einstein's gravitational field, that if the gravitational field deflects lights emitted by celestial objects, then the stars should dance.

His comment was: Einstein's general theory of relativity should not be too involved in the details.

This comment was beyond unexpected. I do not know how he could have figured that out with his young head. This should be a sixty or seventy-year-old man possible to comprehend the secret.

This is my first time to ponder this point: Tao that can be described is not universal and eternal Tao. Lao Tzu's five thousand word <Tao Te Ching> hides like a mountain behind thick cloud and fog, abstract, profound, fuzzy, and high on hype and occult secrets because the Tao is highly abstract and does not rigidly stick to the detail of specific factors, thereby opening the door to all the wonderful, different perceptions of people, even leading to the door of eternal good fortune.

Einstein's theory in detail has left a variety of flaws, so this little guy, like me, can stand here and give my opinion on him.

But high on Lao Tzu, whose Tao has been on ever comprehend, and from one state to another to comprehend a higher realm, where people have no way to see its flaws and give opinion on the flaws.

Few people have ever tried to refute Lao Tzu, but only comprehend his theory further insight, endless.

But Einstein's theory has never seen a lack of criticism, although mainstream science rejects such criticism, but the mainstream scientific community has never answered, and never will answer, non-mainstream scientific community criticism and questioning on relativity theory.

In fact, this principle can also be extended to social life.

In a country, if the ruler cares only about the routine life of firewood, salt, and oil, and leaves the ideological spiritual thought to those in the religious sectors who only talk about the spirit, heaven, hell, afterlife, etc. - illusory, but everyone needs to face them in the days after death, and there is always, after death, the reward and punishment of heaven or hell - the country is always harmonious.

And when the ruler of a country not only manages temporal affairs but also those of the spirit, the social harmony in this country is not likely to improve.

If the management of the spirit also excludes the afterlife: heaven and hell, etc., that will lead to deletion of public behavior norms that come from the soul.

Using only the law, it is impossible to constrain people's selfish desires in every second.

Modern society is a society dominated by capital. The nature of capital is to stimulate people's selfish nature to do its utmost to pursue profit. Without the constraint of heaven and hell, the pursuit of profit in the heart of the little people can reach the point of madness, with benefits such as food poisoning. More money, perhaps, does not make things better; it may even make things worse...

These are the questions my discussion with Johnny started on the street in Amsterdam.

The following is our discussion results on the celestial light propagation, now presented to you in this chapter.

When the conversation turns back, please think about a few questions:

When observing celestial objects with a telescope, is the celestial light propagation speed of the obtained image the same as the laboratory measured light speed?

What are the differences between them?

Such differences can affect the results of analysis of observations to what extent?

This chapter has more junior-high level algebraic equations than the previous chapters - the content needs them - and also, let's together slowly learn to express and apply mathematics to our ideas in a simple way. Of course, readers can jump over the mathematics and just read the text.

To consider the time of processing information for a telescope when calculating the propagation time of a celestial observation image

In modern astronomical observations, almost all of the information from distant galaxies and celestial objects must be received through a variety of telescopes and then processed by computers to obtain observations. These results are presented in the form of basic digital data or images.

The resulting image is an observation obtained after treatment of the telescope's effective astronomical information output.

Since the photons emitted by distant objects must be (all observing instruments represented herein by telescope) perceived through the telescope, we basically cannot directly measure the speed of celestial images. So, while we can accurately measure in the laboratory the speed of a single photon, or a group of photons, emitted based on the time received, we cannot directly and accurately measure the final time the light emitted by celestial objects is received because of the process by which the information is obtained after the telescope receives it.

The figure below shows, after the photons emitted by the celestial object are received and processed by the telescope, the time required for image output, or for the information to become useful, and photon propagation time to astronomical telescopes is not the same. The former is calculated from the observed image speed, while the latter is simply calculated by the moving speed of the photon transmission. The former includes the latter. The latter information is not used by the observer, but we mistakenly believe that a message has arrived.

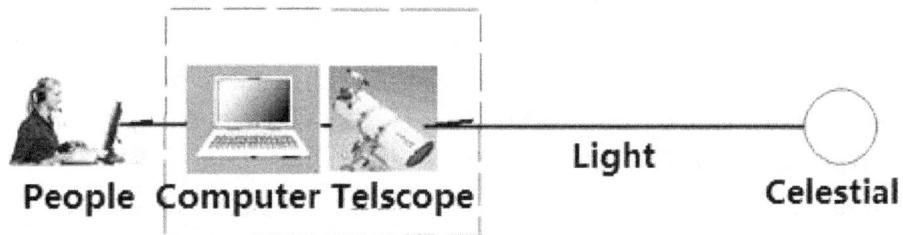

Figure 7.2 right to left. The light from the celestial object arrives at the telescope, is sent as information to the computer, which presents the final output image data (expressed in this image). So far in the analysis and celestial light propagation velocity related information, People never considered the telescope and computer as time consuming. But, in fact, the information must be converted before it is really "seen." The dashed box around the telescope and computer indicate that consume time in the analysis of light propagation speed from a celestial object.

We can describe this situation using the example of a courier. The courier company sends out a parcel for the customer, and 24 hours later, the customer receives the parcel. The delivery event has ended, and the courier company's work has been completed. But the client has not yet gotten the desired product as it is still wrapped. This stage is equivalent to the photon from the celestial object having arrived at the telescope, but the photon arrival information has not been extracted into an image or other useful information to be stored or used. To calculate the image information transmission time, one has to consider the information extraction time.

If the customer immediately opens the package after receiving it, then the process consumes a relatively short time, perhaps only a few seconds, and then the customer really got what he wanted: the desired product. The time it takes for the customer to open the parcel is equivalent to the time it takes for the information of the photon to be consumed by the observer. Similarly, any number of photons reach the telescope, if it reaches the minimum number of responses required for adequate energy telescope required, telescope immediately in a very short time to process and output images, so they end up computing time required to receive the observed image. The telescope information processing time can be very short, but the time is not zero. The telescope is useful for observing any celestial body in the sky or when the available information is consumed.

But if the customer received the parcel and for various reasons cannot immediately open the package, the time it takes to get the desired item inside is prolonged. This prolonged period of time is uncertain.

Similarly, when processing telescope optical information from afar, the observer will encounter the same delay because the information was too weak to make the telescope immediately respond to the analysis, and it takes time to wait for the arrival of more photons. Therefore, the time it takes to obtain the information image is extended an uncertain amount.

According to our hidden celestial chapter, since the distance between the light source and the observer results in attenuation of the light, when approaching the limited radius of an observable celestial body, the light intensity will definitely attenuate to below the minimum energy required for the desired response of the telescope, and the time needed to extract the information that came with the photon is extended, and thus the required time to observe the images of distant objects will also be indefinitely extended.

In the old paper, I considered just a group of photons emitted from celestial bodies, which travel through space to reach the telescopes. This is what I call transmission time in the old paper. This time, coupled with the telescope processing time where these photons arrive and are followed up by other batches of photons and the time observing the final output image, is the real time cost of an astronomical observation image.

When we calculate the speed of astronomical observation of an image, we need to include not only photon travel time, but also output image processing time, which, so far, is not included into the telescope astronomical observation times. It is necessary to consider not just the photon, but also the telescope response time that previously was not taken into account when calculating the speed of light.

Adding this time of forming an output image into the calculation formula brings an unexpected different conclusion that is very important.

It is reasonable and also necessary to consider together the photon travel time and the time it takes telescopes to form the output image because if you want to calculate the speed of the image of celestial emission it must be calculated after the image has been there and after the image of these two time factors has existed.

In modern astronomical observations, when the results do not include the use of integration time, it puts them directly into redshifts. In fact, the more distant the celestial bodies, the feebler the light intensity transmitted to the telescope, the greater the telescope observation integration time required, and the greater the redshift.

In reality, Hubble, for example, uses telescopes to measure the redshifts of objects at distant distances. But he didn't consider how long the telescope handles the photons it receives. This leads to great doubt about his redshift theory and gives us an opportunity to explain the origin of the red shift of celestial bodies from a new perspective.

Is there a possibility that redshift is due to the integration time required for the telescope? Because the Hubble telescope does not consider the time required for integration, can we reconsider its findings and conclusions? Maybe the cause of the Hubble redshift law is not large scale celestial bodies moving away, but a response phenomenon the telescope itself in treating a weak input information.

We will continue to discuss all of this and related issues in more detail later, and go through the design of experiments needed to prove this point.

TELESCOPE RESPONSE TIME

A telescope a few meters in diameter is, of course, not like lab equipment and will respond immediately to the arrival of one photon.

Like in hidden celestial object chapter discussed, the photon processed by the nervous system of the eye must take place after a certain amount of photons enter the eye. Otherwise, there will be too much noise interference. The input requirements of the number of the photons entering the telescope must reach a number large enough that it is at least comparable to the level of uncontrollable noise, so as to overcome the uncontrollable noise and make it possible to analyze the data.

So, for any telescope, after there are a certain number of photons in the input, then these photons will be processed, and the computer will send its output, clean up the storage, and be ready to begin accumulating photons for a new image.

From a professional point of view, this process can be easily explained as follows:

A modern telescope used to obtain celestial images has a storage buffer for arriving photons. The buffer is like a three-dimensional matrix in which the two-dimensional plane corresponds to the observation area of celestial objects, and each point on the plane corresponding to the third dimension is used to record the photons entering the point during the observation period. After the telescope receives an arriving photon, the photon will be stored in accordance with its corresponding storage location in the third dimension of the matrix corresponding to the recorder, the queue being in chronological order. After a certain time, the matrix is filled to a certain extent, and all the photons are output to a computer for processing.

If, in a short time, a very large number of photons (when the light emitted by the celestial object is relatively strong or the distance between is not particularly great) arrive at the telescope, then the counter matrix is filled right away, and the information is sent immediately to be handled by the computer.

But if the celestial object is located at a distance near its relative observable limited radius, the number of photons arriving will be very few, and the telescope will need to wait a long time to be able to receive sufficient photons to send information to a computer. The farther the distance is, the longer the waiting time.

Let's analyze the modern telescope equipped with CCD camera as an example. It works as follows: during the exposure or integrated process, when the photons emitted from a celestial object arrive at the sensor, it produces a response of electrons. The electrons are stored into a potential well until the end of the integration or exposure process. Only then will the computer generate corresponding images.

In fact, the whole thesis of this chapter is to accurately describe the above process in the form of mathematics.

Here we now look and see how to abstract the above description into scientific papers. If you do not like math symbols, you can skip the mathematics. But I suggest that, if you are a student, it is best to read with patience. The first of these are simple; 99 percent only involve junior-high mathematics. More important is that here we can flexibly apply what we have learned from knowledge of books; this will be a lifetime benefit. If you want to read on mathematics, then take a piece of paper and a pen, and write down the important symbolic significance it represents, learn words like mind, like down, what will be the final crystal clear. Several of my writer friends who are old, with pure liberal arts backgrounds, are also be able to do these plainly; they just need to be a little more patient to do it.

Suppose a telescope TEL needs at least per unit area Q (TEL) intensity of photons to produce a response. If the total number of photons that reach the current batch of N_{FULL} meet photon number per unit area that is light intensity $I_{FULL} \geq Q$ (TEL), we define the full light of the telescope response time **FLIRT (TEL) (Full Light Intensity Responding Time)** as follows:

$$\textbf{FLIRT (TEL)} = \textbf{T}_{\textbf{FULL}} \textbf{(TEL, N}_{\textbf{FULL}}\textbf{), I}_{\textbf{FULL}} \geq \textbf{Q (TEL)} \quad (1)$$

T_{FULL} is the complete time the telescope TEL used to process the I_{FULL} required information. Here, when the arriving light intensity is equal to or greater than the information Q (TEL), the information travels from the telescope to the output (usually an image) after the total time required for full-on response, defined as the time FLIRT (TEL). This time is only related with the telescope.

Under laboratory conditions, the full intensity response time FLIRT (TEL) of the photon receiver can be infinitely close to zero. But for an astronomical telescope, when the input light intensity to be treated is very weak, it often needs to accumulate in order to form effective information. This is the information after integration (integrate) period of time. The full light intensity response time FLIRT (TEL) is usually relatively small, but the total we have is FLIRT (TEL) > 0.

In popular terms: the light shines on the lens of an advanced electronic telescope, but the telescope is not immediately able to pass on the information it receives. Like the human eye, the telescope needs to receive enough photons in order to respond to the incoming photons for processing. We need to consider how long the telescope will be required to be able to react.

When the light is very strong, as soon as the light irradiates onto the telescope, it reaches the number of photons required by the telescope within a short time period, so the telescope can react almost

immediately to the light, processing it and then outputting the image. Although the completion of these steps is very quick and the time is short, it still takes some time. One specific example of this is when you focus the telescope at the Sun.

The purpose of our research is to take these nuances and preferably come to find out the law. We want to find out the rules so that this is the first scientific representation of this phenomenon, then we can contact the other relevant factors to find the relevant law f abstract and concrete-related law f (factors) - recall that after the introduction of the preparatory know how. Here, the term "full light intensity response time" is used to describe the case of the telescope with sufficient light to immediately react.

This is just a simple case in which there are more complex points, analyzed one by one below.

Considering distance attenuation effects while compute the time of observed image

The above is from the point of view of the telescope. Also, consider further, what if the light is very weak? The telescope cannot immediately respond. And what is the reason that the celestial light becomes faint? Of course, it is the distance between it and the telescope, and the celestial light intensity.

To compute the time required for astronomical observation of an image, one needs to consider the influences of the distance attenuation effects of celestial batch emitted photon statistics velocity

So we begin to analyze the law that the intensity of light from a celestial object attenuates as it propagates through the usual lengthy process, and combine the analysis with the observational capability of telescopes to achieve the ultimate target of our study.

When the position of the telescope is near the relative observable limited radius defined in the hidden celestial object chapter, because the light coming from celestial bodies gradually becomes very weak until it cannot be observed, combined with the way the modern electronic telescope works, in such observations the overall speed of the image received will change significantly, and this change in the theoretical cosmology has not yet been reasonably considered. In the literature, there is no consideration of the intensity of light emitted by celestial attenuation with distance, the changes in the sensitivity of the telescope receiving the information, and the impact on overall light speed of information dissemination. To correctly distinguish laboratory light speed from the statistically observed light image will change a lot of results obtained from theory analysis.

The distance attenuation effect on the correct analysis of observed image speed is an important factor that cannot be ignored.

Celestial object emitting batch quantities of light

When the celestial object emits light, it is clear that in a very short time, only a limited number of photons are emitted because even all the celestial itself into photons emitted out, only a relatively large number of photons. This situation is illustrated in Figure 7.3.

In Figure 7.3, the celestial object continuously emits photons. In order to facilitate research, we can artificially divide a continuous time into the same minimum time interval: Δt_i is used to represent any one such time interval. In the figure, the celestial object sends out N_i photons per unit area to the observation instruments for every tiny interval Δt_i.

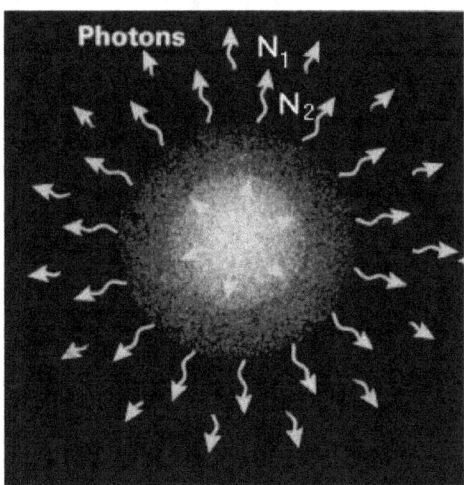

Figure 7.3 Schematic of star emitting batches of photons

The subscript "i" stands for "arbitrary," and i = 1 represents a first time interval t_1, or the first photon N_1. Note that simply for convenience, we used here the number of photon quantities "per unit area," so that later, when writing a formula, we don't have to put "4π" in the equation.

In mathematics, the terms "arbitrary" or "anyone from a lot of" are often, as above, represented with subscripted letters (for example, i or j), and then it's explained that this letter represents any of many (for example, i = 1, 2 ...).

We must master this kind of expression of "infinity," or "a lot." This is the dynamic analysis of higher mathematics, the key knowledge point when deterministic analysis of elementary mathematics transitions to uncertain thinking.

THE ROLE OF DIFFERENT REGIONAL ZONES OF BATCHES OF PHOTONS

We are still going to do the same as before: first represent with simple mathematics, then give the language to describe it.

Suppose we denote the emitting celestial observations by C; TEL indicates telescope; P is any location point; P_{TEL} indicates telescope TEL located at point P; N_i represents the number of all the photons in time i sent from celestial C; $D(C, P_{TEL})$ is located at point P with distance between the telescope TEL and C; $I(P_{TEL}, N_i)$ is the light intensity per unit area of N_i irradiated photons emitted from C to the telescope TEL located at P. At the same time, we express relative observable limited radius as RLOR.

Compile a table in order to easy refer to:

Table 7.1

Symbol	Meaning
$D(C, P_{TEL})$	Distance located at the point P and celestial telescope TEL between C
C	observing celestial object
$I(P_{TRL}, N_i)$	photon per unit area light intensity $I(P_{TEL}, N_i)$ C sent to the telescope TEL at point P
RLOR	Relative Limited Observable Radius
N_i	The N photon of i batch emitted from C
P	represents an arbitrary point
P_{TEL}	telescope TEL located at point P
TEL	telescope

According to basic physics, light intensity varies with distance:

$$I(P_{TEL}, N_0) = \frac{N_0}{4\pi D^2(C, P_{TEL})}$$

Simple math tells us that the limit

$$\lim_{D(C, P_{TEL}) \to \infty} \frac{N_0}{4\pi D^2(C, P_{TEL})} = 0$$

Therefore, when the emitted light intensity C is large enough, we can always find on the way of light propagation so that we put this point named P_{CRIT}, so as to satisfy the following formula from C to P_{CRIT} where this light intensity:

$$I(P_{CRIT}, N_0) = \frac{N_0}{4\pi D^2(C, P_{CRIT})} = Q(TEL)$$

We call P$_{CRIT}$ the **Saturated Response Points**.

We call the point where RLOR is located the **Relative Observable End Point**.

Thus, the following figure shows the photons sent from C at the same photon batch N$_0$ at the same time along the transmission way. In Figure 7.4 the P$_{CRIT}$ interval from C to P$_{CRIT}$ is called the **Saturated Zone**; the interval from P$_{CRIT}$ to RLOR (C) is called the **Distance Hysteresis Zone**; beyond the RLOR (C) is the **Hidden Zone**.

The above derivation is based on that starlight attenuation with distance, and the limited ability of the telescope that can not handle infinitely attenuating weak light.. Thus, it is possible to find a cut-off point near enough to the telescope where starlight, although attenuating with distance, still has sufficient strength so that the telescope can respond immediately and so that the response time is the shortest full intensity response time. This point is called the **Saturation Response Point**, and this area is called the **Saturated Zone**.

And beyond this point, the telescope needs to wait for the arrival of a subsequent photon superposition response to accumulate sufficient light intensity to meet the needs of the telescope, thus lengthening the response time lag. This area is called the **Hysteresis Zone**.

With increasing distance from the Hysteresis Zone, the lag takes more and more time until it reaches the point that the light from the star is so weak that, no matter how much light accumulates, the telescope cannot respond to the lights, and that means the telescope no longer sees this star. This point is called the **Relative Observable End Point**. From this point to C is the relative observable radius discussed in hidden celestial object chapter.

Beyond this point, across greater distances, the telescope never sees the star, so this area is called the **Celestial Hidden Zone**.

Specifically, according to Figure 7.4, we analyzed two points (saturation response point, relative observable end point) and the three regions split by them (saturation region, hysteresis region, celestial hidden area).

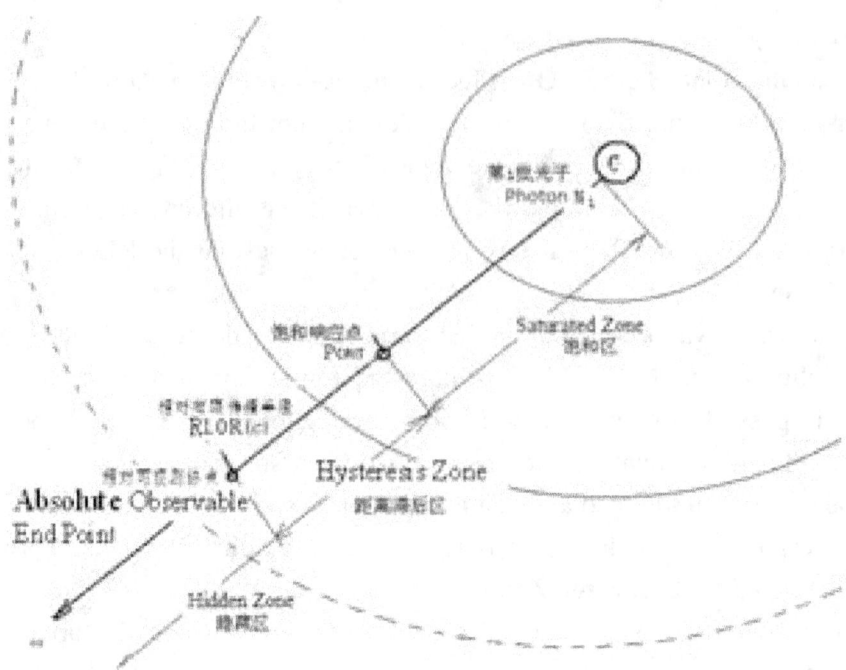

Figure 7.4 The damping effect so that photons emitted by celestial C with different distances between the telescopes will produce different effects, which is divided into the Saturated Zone, the Distance Shifted Zone, and the Hidden Zone.

Saturated zone

Suppose at time T_0 celestial C emitted N_0 photons with speed V_{EMIT} over a distance D (C, P_{TEL}), and at time T_1 arrived at P_{TEL} where the telescope is located inside the Saturated Zone. Inside the Saturated Zone, I (P_{TEL}, N_0) > Q (TEL), so the telescope always can respond in this zone in the FLIRT (TEL) time.

Therefore, the light intensity that the celestial C sent to the telescope I (P_{TEL}, N_0) is calculated as follows:

$$I\ (P_{TEL},\ N_0) = \frac{N_0}{4\pi D^2(C, P_{TEL})} \geq Q\ (TEL),$$

When $0 < D(C, P_{TEL}) \leq D(C, P_{CRIT})$

Noting the above formula is defined to be used inside the Saturated Zone, we know the above calculation is actually calculation of full light intensity and the time FLIRT (TEL).

In the case where other factors are not considered, C emits N_0 at time T_0, which arrives at TEL at time T_1, and the transmission time is $T(C, P_{TEL}) = (T_1 - T_0)$. In such a case, V_{EMIT} is the speed of N_0 that is the first batch of photons C launched, namely the speed of light Einstein defined.

However, at time T_1, we still cannot perceive the observed image OI generated by N_0. We have to wait until the telescope finishes processing and obtains the output image OI. Only then do we know that we received the N_0, and this time is the full light intensity response time FLIRT (TEL). So if we define the total time it takes for N_0 to depart from C, arrive at TEL, be handled by the computer, and be output as an image as TOI (Time of OI), then, in the Saturated Zone:

$$\textbf{TOI}(C, P_{TEL}) = \text{FLIRT}(TEL) + T(C, P_{TEL}),$$
$$\text{when } D(C, P_{CRIT}) \geq D(C, P_{TEL}) > 0$$

Accordingly, the Speed of Observed Image (SOI) of the output image can be simply calculated:

$$\text{SOI}(C, P_{TEL}) = D(C, P_{TEL}) / (T(C, P_{TEL}) + \text{FLIRT}(TEL))$$
$$= D(C, P_{TEL}) / [T(C, P_{TEL})(1 + \text{FLIRT}(TEL) / T(C, P_{TEL}))];$$
$$\text{When } 0 < D(C, P_{TEL}) \leq D(C, P_{CRIT})$$

Let $Z_0 = \textbf{FLIRT (TEL)} / \textbf{T (C, P}_{TEL}\textbf{)}$, where Z_0 is **Full Light Intensity Hysteresis**. Such a formula can be written as:

$$\text{SOI}(C, P_{TEL}) = D(C, P_{TEL}) / [T(C, P_{TEL})(1 + Z_0)]$$

Which can also be written in this way:

$$\text{SOI}(C, P_{TEL}) = V_{EMIT} / (1 + Z_0)$$
$$\text{When } Z_0 = \text{FLIRT}(TEL) / T(C, P_{TEL}),$$
$$V_{EMIT} = D(C, P_{TEL}) / T(C, P_{TEL}),$$
$$0 < D(C, P_{TEL}) \leq D(C, P_{CRIT})$$

When $D(C, P_{TEL})$ is very small, like the distance between the Sun and the telescope, Z_0 is important; when $D(C, P_{TEL})$ is very large, Z_0 is negligible.

The above derivation looks like very complicated, but in fact, only uses elementary algebra. It is not difficult but is tedious and cumbersome. It just requires patience. Here, briefly, is a summary in non-mathematical language:

Assuming TOI is the total time for which light of celestial C is received, and D (C, P_{TEL}) the distance between telescope and the celestial C is known, we can calculate the speed of the output image celestial SOI. This assumed total time in the Saturation Zone actually consists of two parts: one of the time portions is the time the light started from C and arrived at the telescope, which is calculated by the distance divided by light speed, since this light speed is the usual one from the lab; the other time portion is the time the telescope uses for processing strong enough received information.

The speed of image SOI (C, P_{TEL}) is converted into the form containing Z so that it can be used later to compare with the Doppler redshift formula in the same form.

The mathematical technique used here is to divide the continuous starlight into infinitely small intervals (indicated by Δ) and then imagine that the telescope receives light in these small intervals, one by one, until the one is received (denoted by subscript i) that the telescope is saturated with received light (indicated by Δ_i) so that the interval in front of it (indicated by Δ_{i-1}) does not have enough photons to make such a telescope in saturation status.

Distance Hysteresis Zone

Now to be discussed is the situation where the distance between the telescope and celestial body is equivalent to the case in Figure 7.4 where the telescope is located at a point between the saturation point and the Relative Observable End Point. In this range, starting from the Saturation Response Point, as the distance increases, the intensity of the light emitted by the celestial object reduces, and the telescope needs to wait longer and longer to respond. This means that, as the distance increases, the telescope requires more time to respond and output information. This area is called the **Distance Hysteresis Zone**.

In this zone, we have

$$I(P_{TEL}, N_0) = \frac{N_0}{4\pi D^2(C, P_{TEL})} < Q(TEL),$$

when $RLOR(C) > D(C, P_{TEL}) \geq D(C, P_{CRIT})$

In the case of non-full light intensity, the telescope will extend its exposure time, or integration time, until it collects or accumulates enough photons. Then it will convert the cumulative information into an output image.

Since C continuously emit photons, the time can be partitioned into numerous tiny time intervals Δt_i. The number of photons emitted in this range is N_i. So, we observe at the start time interval Δt_0 emitted N_0 photons. Due to fading distance, when the distance is inside the hysteresis zone, I (P_{TEL}, N_0) < Q (TEL), so that the telescope cannot respond immediately. Thus, the telescope receives and stores photon arrival N_0. This is followed by the arrival of the N_1 batch of photons. If I (P_{TEL}, $N_0 + N_1$) > Q (TEL), the telescope begins response FLIRT (TEL) followed by output image OI; otherwise, N_1 batch photons continue to be stored, and the telescope waits for the next batch of photons.

This situation can be represented by Figure 7.5 of the improved Huygens advancing wave peak.

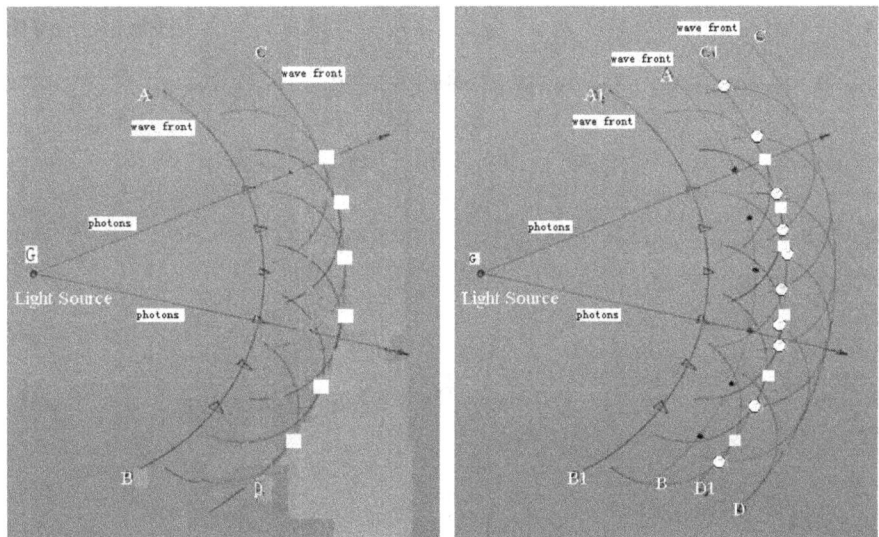

Figure 7.5 Huygens advancing wave peak overlay

On the left image of Figure 7.5, A-B and C-D represent the two peaks coming one after another, and C-D is at the location of the observing instrument. Six white square points represent the contact surface of the photons and the observing instrument. In the right image, the peak surface C-D has crossed C1-D1 where the observing instrument is located, and the newly arrived crest A-B has just arrived at C1-D1, so C1-D1 not only has A-B at the telescope, with six squares representing photons, but also has C-D left on the C1-D1 residue, with white dots indicate such after a Δt. The crests A-B and C-D on the observation instruments complete the

231

photon superposition.

In general, we have:

SOI (C, P_{TEL}) = D(C, P_{TEL}) / (FLIRT (TEL) + T(C, P_{TEL}) + $\sum_{i=0}^{j} \Delta t_i$)

When: $\sum_{i=0}^{j-1} N_i / 4\pi \ P_{TEL}^2 < Q(TEL)$,

$\sum_{i=0}^{j} N_i / 4\pi \ P_{TEL}^2 \geq Q(TEL)$,

LPR(C) $\geq$ D(C, P_{TEL}) > D(C, P_{CRIT})

RLOR(C) $\geq$ D(C, P_{TEL}) > D(C, P_{CRIT})

In the above formula, let $Z_1 = \sum_{i=0}^{j} \Delta t_i$ / T(C, P_{TEL}). This is the **Distance Hysteresis Parameter** of the output image, and the above equation can be written as:

SOI (C, P_{TEL}) = D(C, P_{TEL}) / [T(C, P_{TEL}) (1 + Z_0 + Z_1)]

As well as:

SOI (C, P_{TEL}) = V_{EMIT} / (1 + Z_0 + Z_1),
V_{EMIT} = D(C, P_{TEL}) / T(C, P_{TEL})

So that $Z = Z_0 + Z_1$ is the **Distance Hysteresis Parameter** of the output image, and finally we get:

SOI (C, P_{TEL}) = V_{EMIT} / (1 + Z)

The above derivation is mainly in order to calculate the distance hysteresis parameter.

The principle is based on the idea that the previous batch of photons arriving at the telescope was not enough to stimulate the output image, so that it had to wait for the arrival of a group of one or more subsequent photon batches (here we have to deal with an uncertain concept). Add the elapsed times all together, and we get the total time of all photon batches. The ratio of the total time and the time of the normal light transmission from the celestial object to the telescope is what we seek for the distance lag parameter.

Invisible Zone

If a distance between the telescope and the observing celestial exceeds the distance between the telescope and the relative observable end point, the telescope never sees this celestial object, and that is the "**Invisible Zone.**"

If the $D(C, P_{TEL}) > RLOR(C)$, then C is an invisible celestial, and $SOI(C, P_{TEL})$ no longer exists. Which is:

$$SOI(C, P_{TEL}) = N/A, \text{ when } D(C, P_{TEL}) > RLOR(C)$$

To sum up

To summarize the above analysis, we can get SOI (the Speed of the Observed Image) that is obtained by the telescopes TEL located at P_{TEL} at a distance $D(C, P_{TEL})$ by using the equations in the following table:

Table 7.2. Speed of the observed image of a celestial object in different areas:

SOI =	Saturated zone $V_{EMIT} / (1 + Z_0)$	when $Z_0 = FLIRT(TEL) / T(C, P_{TEL})$, $D(C, P_{CRIT}) \geq D(C, P_{TEL}) > 0$, $V_{EMIT} = D(C, P_{TEL}) / T(C, P_{TEL})$	(1.1)
	Distance hysteresis Zone $V_{EMIT} / (1 + Z)$	when $LPR(C) > D(C, P_{TEL}) \geq D(C, P_{CRIT})$ $Z = Z_0 + Z_1 + Z_2$ $Z_0 = FLIRT(TEL) / T(C, P_{TEL})$, $Z_1 = \sum_{i=0}^{j} \Delta t_i / T(C, P_{TEL})$, $\sum_{i=0}^{j-1} N_i / 4\pi P_{TEL}^2 < Q(TEL)$, $\sum_{i=0}^{j} N_i / 4\pi P_{TEL}^2 \geq Q(TEL)$.	(1.2)
	Invisible Zone N/A	when $D(C, P_{TEL}) > RLOR(C)$	(1.3)

Note:	
SOI	Speed of Observed Image
V_{EMIT}	Emitted photon speed, $V_{EMIT} = D(C, P_{TEL}) / T(C, P_{TEL})$
Q (TEL)	The minimum required number of photons in excitation to cause a response of the telescope TEL
P_{CRIT}	Saturation response point, the telescope located at this point to meet $N_0 / 4\pi D^2 (C, P_{CRIT}) = Q (TEL)$
P_{TEL}	Telescope TEL located at this point
FLIRT(TEL)	Time of Full Light Intensity Responding of a telescope TEL when $I(C, N) \geq Q (TEL)$
N_i	Photon number of batch I the celestial object emitted at Δt_i
RLOR(C)	Relative Limited Observe Radius of celestial C
Z	The Hysteresis Parameter $Z = Z_0 + Z_1$
Z_0	Full Light Intensity Hysteresis parameter, $Z_0 = FLIRT(TEL) / T(C, P_{TEL})$
Z_1	The Distance Hysteresis Parameter The Distance Hysteresis Parameter $Z_1 = Z_1 = \sum_{i=0}^{j} \Delta t_i / T(C, P_{TEL})$

The table looks very complicated. In fact, it just has the three different results of the propagation speed of the observed image SOI varies with distance in three different ways. Although with the definition of each symbol and the different conditions all together, it looks very cumbersome.

The following mathematical model summarizes the above discussion. It represents our new interpretation of redshift. It is the Hysteresis parameter of redshift, and is a composition of celestial object observation image lag redshift parameters.

TIME HYSTERESIS REDSHIFT AND DISPLACEMENT REDSHIFT

If the distance between telescope TEL located at P_{TEL} and the observing celestial C has a significant relative movement, that is, the value of D (C, P_{TEL}) is increased, then the change is caused by **Displacement Shift**, represented by Z_2. Note that the displacement shift may be blue shift if the telescope TEL and observed celestial C move closer. As our focus is now on the Big Bang theory of redshift, we will not discuss blue shift here further.

Redshift Hysteresis Parameter of Observed Image without considering blue shift is:

$$Z = Z_0 + Z_1 + Z_2$$

Old astronomy redshift is defined by the speed of the original formula:

$$Z = \frac{V_{EMIT} - V_{TEL}}{V_{TEL}}$$

wherein V_{EMIT} is emitted photon speed,

Wherein V_{EMIT} is emitted photon speed,
V_{TEL} is the observed star speed by TEL

So we have:

$$V_{TEL}(1 + Z) = V_{EMIT}$$

But the formula does not consider the impact of any lag parameters; the causes of redshift Z are simply all attributed to the relative motion of celestial bodies transmitting to and received by the telescope; according to astronomical observation theory, objects redshift solely by the motion of moving away from the telescope and does not take into regard any of the impact we proposed and discussed above in addition to various other external displacement lag parameters.

We believe that the correct approach is to take all the factors that can cause redshift on an astronomical telescope into account. Therefore, (1.2) in Table 7.2 where there are,

$$Z = Z_0 + Z_1 + Z_2$$

Wherein $Z_0 + Z_1$ is called **Time Hysteresis Redshift**, and Z_2 is redshift due to the **Displacement Redshift** between the celestial object and the astronomical telescope.

Doppler redshift theory only works when the distance between the celestial object and the telescope is large enough, usually more than millions of light-years. And we believe that, because of the telescope processing celestial information features, even celestial objects near the saturation region have full light intensity hysteresis redshift parameter Z_0. A large number of astronomical observation data proved this point. For example, the Sun is only 490 light-seconds away from a telescope on Earth, but its redshift was measured. This was discussed earlier in full light intensity Time Hysteresis Redshift. The full light intensity hysteresis redshift also may be measured from other nearby stars.

Experimental Design Feasibility Study I: receiving different intensity acoustic waves from a same acoustic source

This is the simplest of the initial tests to determine the feasibility of our ideas that the massive celestial redshift is not caused by the Doppler movement.

1. Initial test:

Let the car travel at a constant 100 km/h speed each time the test is repeated. As one experimenter drives away, sound waves are sent out of the same intensity, the same frequency, and from the same sound source. Those listening to the sound should block their ears to varying degrees during each repetition of the experiment to see if each time they receive a different volume (non-blocked ear - the sound is very loud - compared to almost completely blocked ear - the sound is weak) and to see if the tone has changed. If there is a significant change in the volume but no significant change in tone, it shows the acoustic wave represented will not change its frequency due to changes in volume. Extend to light (as Doppler did), that is on the earth while receiving waves of celestial bodies with different sensitivities, the sensitivity of the intensity of the received light with different receiving equipment changes, but its frequency does not occur major changes, namely redshift measured value will be no major changes.

I did a little of the above experiment. The results, as I conjectured, are that changing the volume of sound waves does not cause a large change in frequency.

Such experiments are roughly equivalent to a feasibility study. I will continue to make more accurate, precise scientific experiments and leave the record so that others can repeat the experiment.

2. Accurate test that can be repeated

The scientific experiments record accurately and be able to repeat step by step.

Probably we need to record the following:

Time, place, wind, wind direction, participating car models and other parameters, the parameters involved in the original sonic and duties of people, traffic, sound source, each adjustment of the volume of data and the original waveform, and other parameters from the receiving location, each time waveform received. The final report and conclusions should be summarized.

If the experiments are successful, we can be formally go on to the observatory to do observation experiments. This should allow us to apply for funding related to research and to find people interested in working at the observatory.

Experimental Design II: For Celestial Observation Image Redshift Hysteresis Parameter Z

There is no way to properly test the speed of the observed image of a celestial object by experiments on Earth similar to the ones Michelson-Morley made, because there is no way to simulate celestial light after attenuation through the dissemination of many light-years in a laboratory environment. But we can use modern telescopes to simulate and test the observed image speed. We can do the experiment with a telescope as a receiver for a strong light source, blocking out the photon numbers received from the emitting celestial object, starting from a few photons and gradually increasing the number of photons received by the telescope, observing the exact time used in every test from the beginning of photon emission to computer output image. From that, we can determine Z_0, and Z_1. We can also simulate the number of photons and redshift on normal telescopes and do comparisons to determine the correlation value.

For an astronomical observation image lag observed redshift experiment we can use a telescope to observe the Sun. Comprehensively and incrementally block out the Sun's light, and observe the consequent increasing redshift to find out the law. By comparison with other standard celestial bodies, it is possible to separate Z_0, Z_1, and Z_2 from $Z_0 + Z_1 + Z_2$.

This experiment has a great chance of success, citing astronomical observations used in the filter and related K Correction theory,

We know that filtering out some of the light energy will cause the redshift value to increase accordingly. Suppose a spectral redshift object has a value $Z = 1$. Since the filter reduces light intensity, the redshift object will become two elements, namely $Z + 1$. Therefore, luminosity (photometric) measurements due to the use of filters will increase the redshift.

We are now using the same principle to do the experiment. In our experiment, using the color filter that can filter out a variety of light, all the light, the light intensity of different filters, try to simulate light intensity received from distant celestial objects to find out the appropriate law, coming by a standard way to Z0, Z1 and Z2 separately. The corresponding times of the various observations will be different; be careful not to "calibrate" out the useful information while aligning the telescope.

Conclusion

The concept of celestial object observation image speed does not simply consider the single variable of the movement of celestial objects. Instead, it also systematically considers factors such as light attenuation by distance, the fact that, at a distance exceeding the saturation zone, the telescope needs to wait longer so that the effect seems to be the speed of light "slowing" down the processing time before the valid information is received, as well as the displacement redshift caused by the relative motion of celestial bodies. This allows us from a new angle to consider a more comprehensive calculation of speed of the celestial observed image, thereby expanding the main factors causing the redshift of celestial objects and finding the main cause of celestial redshift.

A simple and effective design of our experiments will further prove astronomical observation image of speed and presented herein reasons that caused the celestial objects redshift.

Redshift theory is the basic support of the Big Bang theory. If there is no expansion of the universe from redshift, the Big Bang theory loses its power, and the generally accepted universal framework of the last few decades will collapse.

It would be a catastrophe of modern cosmology!

CHAPTER 8

DANCING ON A SESAME SEED
- A Review of NASA's Whole Sky Map
- On the Unreality of the Images of the Distant Universe

If topography is measured from the results of the study on a sesame seed, and then extended to the whole of the mountains and rivers of the Earth, what happens?

If we were to carry out an investigation of nationwide distribution of haze, of course, it would require the deployment of thousands of observation points nationwide.

A well-known expert, went to the inside and outside Beijing urban, suburban rural, mountainous areas, to collect accurate data points for haze, and then extrapolate from them national estimates - not just Beijing but all of China. Even with a variety of mathematical models, based on data trends in certain directions, draw out the haze national variation map.. Should we say that this is a super-intelligent expert?

Let us assume that there is a specialist in the range of hundreds of kilometers surrounding South Sea Island, on the water, large and small islands, many observation points on the area, and the data collected from these points are analyzed and calculated accordingly. Can this expert claim that it depicts an accurate map of whole world's oceans? Would any ship's captain dare to use this map on an expedition across the Pacific?

Even in the observations with the world's most advanced equipment, the best mathematical models and the best possible results, because of ridiculous observation model, the called national map and world map can only be reduced to jokes!

However, if well-known experts come out with fancy mathematical models that demonstrate the reasonableness of these models, does that make them more believable?

Think about it; do not answer too fast.

If we agree that the above examples are proof that a very small sample cannot stand for much larger system, then please continue reading this chapter.

Over the last decade, NASA has continued to publish a variety of whole-sky maps of the universe. The observation model and data sources is not essentially different from the above pattern; in fact, it is on an even grander scale. The haze national map represents a country with an urban area, from the area in terms of no more than one to a million (1 / 1,000,000). The global water area distribution map of the water area surrounding the South China Sea local to represent of the global, from the area in terms of no more than one to a million (1 / 1,000,000).

But compare those dimensions to NASA's whole sky map of the universe. The ratio of the area in terms of the proportion of the various space of NASA satellite detection range is determined to be more than one to one hundred **trillion** (1 / 10,000,000,000,000)! Should we not examine some of these amazing universe maps?

In scientific observation, observation equipment is very important, the data analysis is very important, but the most important is neither of these. In scientific observations, the most important is that the observe model is scientific and reasonable. Otherwise, you get "garbage in, garbage out," only with more sophisticated instruments, more perfect data, more deceptive, and farther from the truth. And NASA whole sky map depicting observation, from the beginning of the design, is completely in violation of the laws of nature!

So, do NASA cosmic experts dare to so brazenly carry out such non-scientific research?

Of course not!

QUESTIONS ABOUT THE WHOLE SKY MAP OF THE COSMIC MICROWAVE BACKGROUND

I sometimes see news releases from NASA about the whole-sky map of the background microwave radiation, and I always ask myself:

What is the 'whole sky'? Is the sky around the Earth or the entire universe? If it is the whole universe, how can the whole be observed? Where do weak waves come from? Where do they go? How intensity of its source? How strong the wave will be at the place to go?

After I started thinking about the problems of the universe, still do not understand these issues. I assume I own most advanced telescopes better than any in the scientific fiction or beyond any imagination, I am the Superman, but I still cannot think of any way to cross this very great distance moat and get enough test data to paint the whole sky map.

I was looking for answers on NASA's website, so I read some professional literature, and finally seemed to understand.

They have a sublime theory, they have money, authority, and there is strong support for NASA.

But their theory cannot withstand scrutiny; their observations, from the beginning, had problems with regards to how to design the observation model.

We can sum up the questions to the following points thus:

• The receiving space the human receiving the cosmic microwave background is limited in terms of a 'point' range at the Earth。

• The duration time of the human receiving the cosmic microwave background is limited in terms of a short time piece in the history time;

• Is there any way to measure the microwave radiation between distant galaxies, or between the stars? Can such weak microwaves be transmitted millions of light years from their source? NASA considers the intensity of the cosmic microwave background to be less than, and it is basically is the same everywhere in the universe. However, a wave is always in movement, emanating from the light source to the surrounding space. Microwaves in motion that reach the Earth are measured around 3k, which is indicative of the intensity of the wave on the road to getting it to is more than 3k, and the intensity in the way of its departure becomes smaller and smaller. Assuming that this is likely to change in units of one light-year, or hundreds of light-years, what is the point, within a few light seconds of the Earth measured the substantially stable data value? Is it not the same as taking repeated measurements from smaller than sesame space and then extending the spatial data obtained from the sesame to the whole universe? How can NASA experts deny this hypothesis?

• Is the microwave in the space around Individual stars, supernova … just about 3k? How can one say "whole sky"? Recall the relationship between dark and bright spaces discussed in chapter one, are microwave distributed in a dark space is the same as in the light space?

• Light is mobile, space is expanding, and the microwave 'isotropic' is in conflict with NASA Big Bang theory.

• The speed of the space expansion caused by Big Bang compared to the speed of light leads to an incompatible contradiction.

• We should not use Copernicus' law of the universe at will.

• If it is determined that the microwave background is a product of the Big Bang, logically, the microwave background cannot be used as evidence in turn for the Big Bang theory. As the data measured from one point, there is no any basis can extend the data to the whole universe, this is only possible under the Big Bang assumption. Therefore, using the microwave background as evidence to support the Big Bang theory, is trapped a vicious cycle of circular argument.

This last point is explained in the following table is an intriguing example of the control logic of which:

I assume that all car accidents are caused by drunk drivers.	Astronomers assume that all microwaves are caused by the Big Bang.
I have found a good number of road accident caused by drunk drivers.	Astronomers found microwaves of unknown source at a number of points near the Earth
By extension, using the found data as evidence, I conclude that all car accidents are caused by drunk drivers.	By extension, using the found data as evidence, they conclude that all microwaves are caused by the Big Bang.

One by one in detail, let's take a closer look at our views, how do the theory, experts, satellites, and the data that was collected from the Earth, now the microwave evolution into a whole-sky map.

Profile of the whole sky map of the cosmic microwave background

NASA images depict the cosmic microwave background whole-sky map, the result of a broad and sustained use of resources (allegedly, they have spent more than one trillion dollars over fifty years). This is an incredible model of observation.

The discovery of the cosmic background radiation in modern astronomy is very important, as it gave rise to and great evidence of the Big Bang theory, and, together with quasars, pulsars, and interstellar organic molecules, is known as one of the "four great discoveries", Arno Penzias and Robert Woodrow Wilson won the 1978 Nobel Prize in Physics for their discovery of the cosmic microwave background radiation.

Figure 8.1 shows the juxtaposed equipment and different outcomes during the study of the cosmic microwave background. Sequentially from top to bottom: Penzias and Wilson period, COBE period, WMPA period and Plank period.

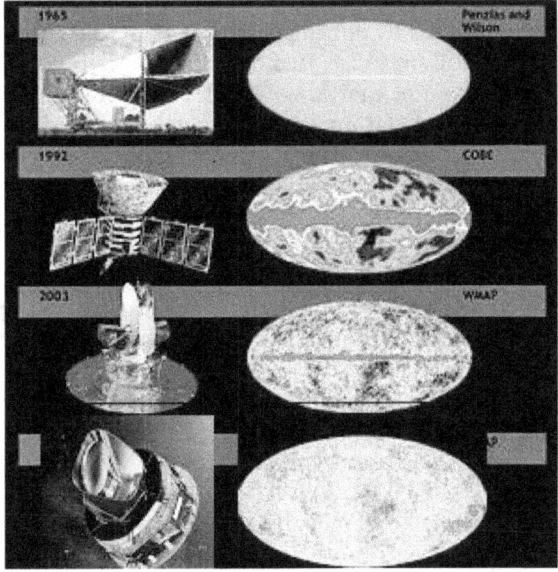

Figure 8.1 Equipment and achievements of microwave background in different periods

Let's look briefly at the theory of the cosmic microwave background and observations of aggregated history. Table 1 shows the cosmic microwave background from theory to observation.

Table 1. Observations of the Cosmic Microwave Background Timetable

---
1941 – Andrew McKellar was attempting to measure the average temperature of the interstellar medium, and used the excitation of CN doublet lines to measure that the "effective temperature of space" (the average bolometric temperature) is about 2.3 K

1946 – George Gamow calculates a temperature of 50 K (assuming a 3-billion year old universe),

1948 – Ralph Alpher and Robert Herman estimate "the temperature in the universe" at 5 K.

1949 – Ralph Alpher and Robert Herman re-re-estimate the temperature at 28 K

1953 – George Gamow estimates 7 K.

1956 – George Gamow estimates 6 K.

1957 – Tigran Shmaonov reports that "the absolute effective temperature of the radioemission background ... is 4±3 K".
1960s – Robert Dicke re-estimates a microwave background radiation temperature of 40 K
1964–65 – Arno Penzias and Robert Woodrow Wilson measure the temperature to be approximately 3 K. Robert Dicke, James Peebles, P. G. Roll, and D. T. Wilkinson interpret this as a signature of the Big Bang.
1966 – Rainer K. Sachs and Arthur M. Wolfe theoretically predict microwave background fluctuation amplitudes created by gravitational potential variations between observers and the last scattering surface (see Sachs-Wolfe effect)
1968 – Martin Rees and Dennis Sciama theoretically predict microwave background fluctuation amplitudes created by photons traversing time-dependent potential wells
1969 – R. A. Sunyaev and Yakov Zel'dovich study the inverse Compton scattering of microwave background photons by hot electrons (see Sunyaev-Zel'dovich effect)
1983 – Researchers from the Cambridge Radio Astronomy Group and the Owens Valley Radio Observatory first detect the Sunyaev-Zel'dovich effect from clusters of galaxies
1983 – RELIKT-1 Soviet CMB anisotropy experiment was launched.
1990 – FIRAS on the Cosmic Background Explorer (COBE) satellite measures the black body form of the CMB spectrum with exquisite precision, and shows that the microwave background has a nearly perfect black-body spectrum and thereby strongly constrains the density of the intergalactic medium.
January 1992 – Scientists that analysed data from the RELIKT-1 report the discovery of anisotropy in the cosmic microwave background at the Moscow astrophysical seminar.
1992 – Scientists that analysed data from COBE DMR report the discovery of anisotropy in the cosmic microwave background.
1995 – The Cosmic Anisotropy Telescope performs the first high-resolution observations of the cosmic microwave background.
1999 – First measurements of acoustic oscillations in the CMB anisotropy angular power spectrum from the TOCO, BOOMERANG, and Maxima Experiments. The BOOMERanG experiment makes higher quality maps at intermediate resolution, and confirms that the universe is "flat".
2002 – Polarization discovered by DASI.
2003 – E-mode polarization spectrum obtained by the CBI. The CBI and the Very Small Array produces yet higher quality maps at high resolution (covering small areas of the sky).
2003 – The WMAP spacecraft produces an even higher quality map at low and intermediate resolution of the whole sky (WMAP provides no high-resolution data, but improves on the intermediate resolution maps from BOOMERanG).
2004 – E-mode polarization spectrum obtained by the CBI.
2004 – The Arcminute Cosmology Bolometer Array Receiver produces a higher quality map of the high-resolution structure not mapped by WMAP.
2005 – The Arcminute Microkelvin Imager and the Sunyaev-Zel'dovich Array begin the first surveys for very high redshift clusters of galaxies using the Sunyaev-Zel'dovich effect.
2005 – Ralph A. Alpher is awarded the National Medal of Science for his groundbreaking work in nucleosynthesis and prediction that the universe expansion leaves behind background radiation, thus providing a model for the Big Bang theory.
2006 – The long-awaited three-year WMAP results are released, confirming previous analysis, correcting several points, and including polarization data.
2006 – Two of COBE's principal investigators, George Smoot and John Mather, received the Nobel Prize in Physics in 2006 for their work on precision measurement of the CMBR.

2006-2011 – Improved measurements from WMAP, new supernova surveys ESSENCE and SNLS, and baryon acoustic oscillations from SDSS and WiggleZ, continue to be consistent with the standard Lambda-CDM model.

2014 – On March 17, 2014, astrophysicists of the BICEP2 collaboration announced the detection of inflationary gravitational waves in the B-mode power spectrum, which if confirmed, would provide clear experimental evidence for the theory of inflation. However, on 19 June 2014, lowered confidence in confirming the cosmic inflation findings was reported.

2015 – On January 30, 2015, the same team of astronomers from BICEP2 withdrew the claim made on the previous year. Based on the combined data of BICEP2 and Planck, the European Space Agency announced that the signal can be entirely attributed to dust in the Milky Way.

The most used probe is the Wilkinson Microwave Anisotropy Probe (WMAP), which obtains huge amounts by its detection of the cosmic microwave background batches. Here is a typical whole-sky map:

With the US satellite launch, other countries of the world with the ability, for example, the Russian, Europe, have also launched satellites. The whole-sky map has been expanded to include objects from the cosmic microwave background radiation to other ray wavelengths, such as infrared all-sky map of the universe and so on.

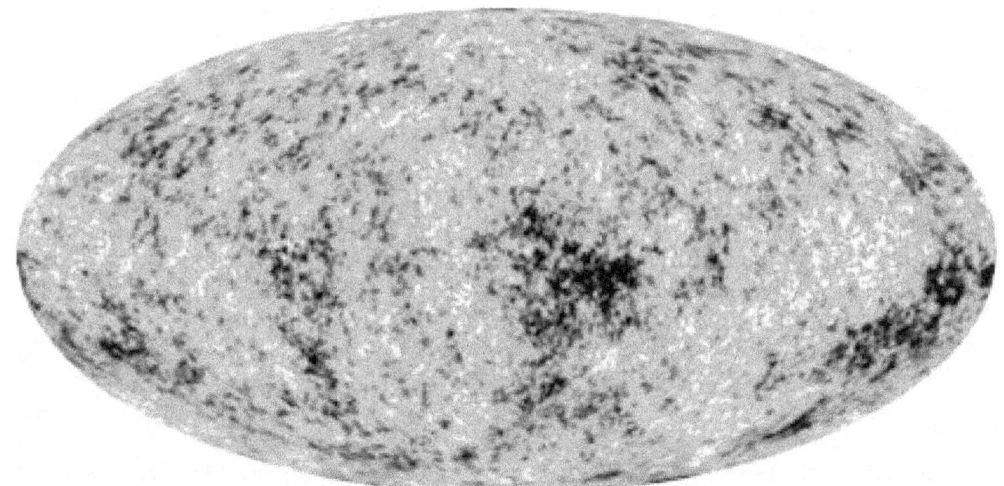

Figure 8.2 the cosmic microwave background whole sky map drawn according to WMAP detected data

Self-circular argument of NASA Big Bang Theory

Cosmic microwave background is one of the bases for proof of the Big Bang theory. Based on the above descriptions, the process of the cosmic microwave background used to prove the Big Bang theory can be summarized by the following Figure 8.3.

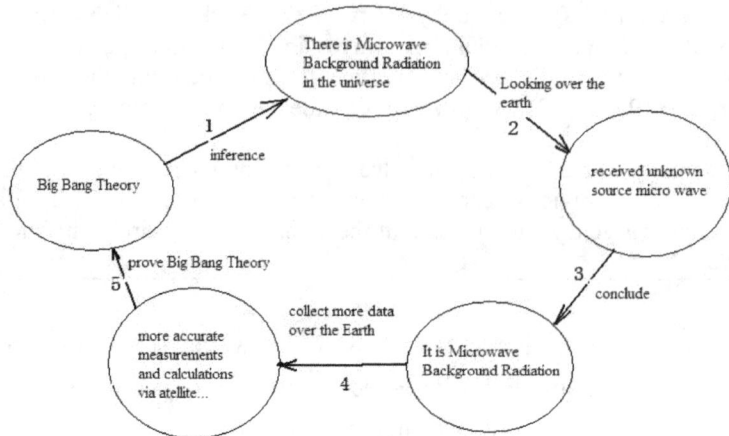

Figure 8.3 Proof of the Big Bang theory using microwave background

Step 1: From the Big Bang theory, we infer universe microwave background radiation, and then

Step 2: Search over the Earth for the cosmic microwave radiation, and found it.

Step 3: Made conclusion that it was microwave background; then

Step 4: The satellite carries sophisticated instruments outside the atmosphere above the Earth and receives data for more than twenty years;

Step 5: The results of these data prove the Big Bang theory.

This is a perfect circular argument!

The problem is that all the observations can be said to have left the Earth!

How big the universe? Let's temporarily accept NASA's estimate of 13.7 billion light years.

How big is the microwave background satellite orbit probe? Plank satellite's orbit is about 5 light-seconds, equivalent to about one hundred thousandth of a light-year.

The radius of the universe is 7 billion light years (assuming that the Earth is at the center). The ration of the orbit radius of microwave background satellite detector to the radius of the universe is 1:7 million billion.

If the circumference of the Earth at the equator, which in reality is 40,075 km long, were the size of a sesame seed, we assume it is 1 cm in diameter, the size of sesame seeds is one of the points relative to the earth 400 million times the maximum circumference!

For example, take the circumference of sesame seed relative to the circumference of the Earth. That relationship is far, far less than the proportion of the microwave background detecting satellite orbit relative to universe's radius (70 million-billionth to more than 400 million equals 17,500,000: 1).

What if we now take a sesame seed outside and imagine someone holding a precision instrument measures on the sesame seeds, and from that we make a lot of mathematical models, and then declare the result of those measures on behalf of the whole Earth? How reliable would that be?

Here's another way of looking at it. People in China used to use *feng shui*, a Chinese traditional form of geomancy, to predict which mountains contained gold. If a mountains really did have gold inside, then we would assume that *feng shui* is a valid method of locating gold deposits Because in the past there was no formal understanding of geology, no other theory that could explain how to find gold, ancient Chinese people had no reason not to trust *feng shui*. This is similar to the drunk driving accident example in previous discussion of the comparison table.

We will now go into more detail with the basic laws of physics related to anatomy at the main arguments of the cosmic microwave background.

THE LIGHT WAVES ARBITRARY PASSING THROUGH A POINT P IN THE SPACE

As preliminaries, we will first discuss a problem: in the absence of any space objects at the point P in the empty space, what things possibly would be exist?

In fact, my 2005 book "谁有权谈论宇宙 *Who Should Talk about Cosmos*", contains a very detailed answer to this question. Readers can refer to the section 'Superposition of Light' on page 71 of the second chapter 'Naked Eyes See the Universe.' It is directly related to this section. Here we will discuss it a little deeper.

At any point P where the space absence of celestial bodies, there are myriad lights emitted from a variety of different celestial bodies through that point. For any celestial body, in terms of years of age, that is greater than the value of its distance to the point P in light-year unit, its light waves will be at that point. For example, if a star's age is 10,000 years old, and the distance between it and the Earth is 12,000 light-years, its emitted light can't shine on the Earth. If the star's age is 15,000 years old, then its emitted light will shine on the Earth. But there is one condition that needs to be met: the intensity of the light emitted by celestial bodies after distance attenuation does not decay to zero; or will not decay to below undetectable levels for the receiving instrument at point P.

If a receiver is put at P, such as the receivers on the Plank satellite, all of the information it receives from the universe can only be those near or far from the entire waves trek passing point P. There is no light in the universe that is not moving. Lights from the light source flow diverging, with the distance between the source increases the light intensity decreases.

But it seems the microwave received by NASA satellites is static, and is not flowing.

Starting from the light source, the light ball is growing, and the weaker energy is constantly spread into space. If the observation target is 'X' light-years away, the place where the observer receiving the light from 'X' is historical light of 'X' years ago from the target. Numerous celestial lights pass through point P, so numerous waves are superimposed at point P. The telescope is just a receiver and resolver: it receives all celestial lights, then filters out the celestial lights that do not need to be observed, leaving those lights from the celestial objects we need to observe.

We broadly classified the countless waves passing through point P, and get Figure 8.4. This figure is copied from my first book "Who Should Talk about Cosmos", but here I've added the k6 and k7 two types of lights.

These light waves contain several invisible lights that would make people crazy. Let's think about how many types of light with no source can be measured.

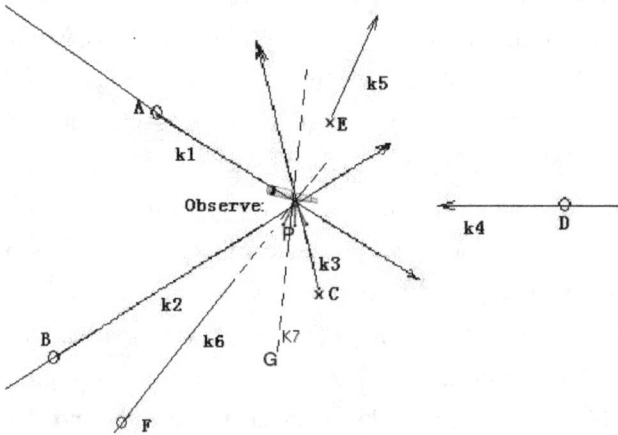

Figure 8.4 various types of light sources, each line represents one type of light source, passing through point P. Please note that each type represents a huge number of similar celestial light objects.

A, B, D, F are represented by a normal celestial circle icon o near the end of a line. Light waves A and B normally pass through the telescope at arbitrary point P. D is different: it's too young, its light (represented by k4) has not yet reached P. D actually exists only in our reasoning. F is trickier, because its light transmission exceeds its saturation response point after some distance, (see Chapter II, Figure 2.4). Because of its light attenuation badly, what is visible to the observer is not continuous light, so it is indicated by a dotted line. It is located in the vicinity of its limited observable radius.

G celestial light of K7 is too attenuated; its light source, while there, cannot be identified. This is a class of hidden objects whose light is very weak, but not non-existent, can be lower than the differential noise from the telescope out at P levels. But if there is a plurality of such residual waves superimposed on hidden objects, it is possible to add these together to achieve a significant energy level.

Celestial bodies C and E, represented by icon X, means the light source does not exist. E only exists in our imagination, and C was discussed in the section of "Seeing is not real - reality is illusory" on page 73 of my first book "Who Should Talk about Cosmos". Suppose k3 represents the value of light-years of the distance from points P to C. Because C disappeared k3 years ago but its light has continued to spread through the point P, it will continue to spread to point P for k3 years! If k3 = 300 million light-years, and C was born four million light years ago 5 million light-years away from Earth, and then C was destroyed 100 million years ago, its light will continue to be visible on Earth for another 300 million years On a smaller scale, Stephen Hawking quipped that since sunlight takes 8 minutes to reach Earth, if the Sun died now, we won't know it for eight minutes

So we can see from Figure 8.4, there are three main types of celestial bodies in the universe represented by F, G and C, and many celestial objects that belong to these three categories: a large number of lights passing through point P whose source cannot be observed.

A DISCUSSION OF THE ISOTROPY OF THE MICROWAVE BACKGROUND WITH INVISIBLE WAVE SOURCE

Figure 8.4 tells us that human beings cannot detect large numbers of light waves from three types of sources F, G and C passing through point P. This large number of unknown sources of light very likely includes so-called cosmic microwave background radiation waves. I remember reading an article from NASA positing with a graph that the cosmic microwave radiation looked as if coming from two distant, non-exist celestial bodies, which is in line with the situation here we are talking about. Hope there is reader willing to help find this article published by expert from NASA. Thanks!

In short, an antenna on earth or satellite in orbit received weak microwaves, and because it could not find the source, it was concluded that it is the residual wave background of universe after the Big Bang, which is not logical; there is no sufficient scientific basis for this statement!

Figure 8.5 is the screen shots of web pages from Lawrence Berkeley Laboratory, one of the four major US research labs, entitled "The Cosmic Microwave Background Radiation".

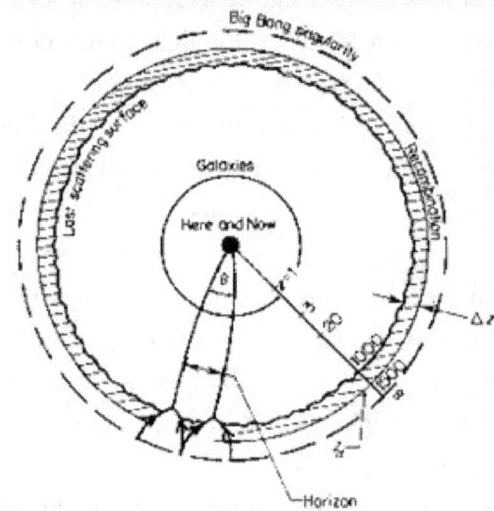

This diagram, centered on the observer (you), shows a representation of the universe where the angle represents the angle of view and the distance (radius) from the center measures both distance and time since light travels at a finite speed.

......

The surface z=1000 is sometimes called the cosmic photosphere, in comparison with the photosphere (apparent surface) of the Sun. **It is the surface from which the cosmic background photons last scattered before coming to us.**

The light coming from this cosmic photosphere (surface of last scattering) can be used to make an image of the early Universe......

However at the time, cosmologists, having very little data, **fell back on the Cosmological Principle: which states that the universe, on the average, looks the same from any point. It is motivated by the Copernican argument that the Earth is not in a central, preferred position. If the universe is locally isotropic, as viewed from any point, it is also uniform. So the cosmological principle states that the universe is approximately isotropic and homogeneous, as viewed by any observer at local rest...**

Given this qualification checked in limited regions by small angular scale observations, **any attempt to interpret the origin of the CMB as due to present astrophysical phenomena (i.e. stars, radio galaxies, etc.) is discredited. Therefore, the only satisfactory explanation for the existence of the CMB lies in the physics of the early Universe.**

Figure 8.5, "The Cosmic Microwave Background Radiation" of web pages (text only interception of the relevant part of the http://aether.lbl.gov/www/science/cmb.html) italics and bolds added by author. Interested readers may go to the Web age in its original (2015-7-12) from the web page of major US Lawrence Berkeley Laboratory

The lowermost bold text of Figure 8.5 says: "any attempt to interpret the origin of the CMB as due to present astrophysical phenomena (i.e. stars, radio galaxies, etc.) is discredited. **Therefore**, the only satisfactory explanation for the existence of the CMB lies in the physics of the early Universe." According to above Figure 8.4, there are three categories of waves without source, thus this "Therefore" assertion is incorrect, because the source is possibly in these three categories of waves.

Because they cannot find the microwave radiation source, they have concluded that the cosmic microwave background radiation wave is the remnants of the Big Bang.

Because there is no other explanation for the redshift, they have concluded that redshift is caused by the expansion of the universe. As to the situation before the explosion and the outside of it after the explosion, it doesn't matter.

Because we cannot see the hidden celestial objects, they have concluded that a given hidden celestial object is dark matter. In fact, there are plenty of these objects that can be seen with the right equipment. Experts sometimes turn a blind eye.

Because we cannot measure a few thousandths or millionths of elementary particle size or weight, photon is defined as being weightless, but just because we can't do it now doesn't mean that it doesn't have weight.

Because did not expect gravitational field is permanent exists and is always there, but rather thought it was instant communication, so defined a curved time space. But why is space-time curved? What is it based on?

...

From the perspective of the light wave propagation in Figure 8.4 we see the result of countless waves passing through points P. Compared to these waves, the satellite observation orbit is nothing. They are light waves superimposed at point P, millions, billions, trillions of times bigger than the satellite observation orbit. These light wave front curvatures are so great, it seems as if they are countless flat light walls passing through point P away in all different directions. If there are many celestial objects, and microwaves from each of them stream towards Earth from all directions, superimposed together, they will become isotropic and seem to be a background wave with no source.

Conversely, the observed isotropic microwave radiation background is negative evidence for the Big Bang theory; that is, after the Big Bang microwave background must not isotropic. From the normal way of thinking, the microwave background is the Big Bang emission center, so the microwave close to the Earth is emitted after the Big Bang expansion from the center outward through the Earth. Did you ever see any explosion will keep its smog always isotropic - which means the smog will never disappear, and universe does not expand?

Paradox between the Big Bang theory itself and the microwave background radiation

Let's read the bold text in the second paragraph of Figure 8.5. It reads, "The surface z=1000 is sometimes called the cosmic photosphere, in comparison with the photosphere (apparent surface) of the Sun. It is the surface from which the cosmic background photons last scattered before coming to us." The meaning of this passage is not so clear. As a precaution, I referred to some documents to determine its meaning is, and this is what I came up with: NASA measured cosmic microwaves from the space opened up by the Big Bang almost outermost aperture, it is labeled the aperture 1000, after some of the changes, and finally reach above the Earth, is received by the antenna of the instruments, and satellite.

We know that any telescope or observer can only passively wait for the arrival of celestial light. The light from most celestial objects is distributed for at least 13.7 billion light-years (the currently accepted size of the universe). And this is only the light emitted by celestial bodies. A whole-sky map is not a whole-sky star map, which describes the distribution of the microwave background all over the space in the universe. Is there any way to be able to measure microwave between galaxies, or stars? Remember, a wave is always in movement, emanating from its source to the surrounding space.

Does microwave propagation need to obey the law of wave propagation? Or, from a thermodynamic point of view, it does not need to obey the law of heat propagation? NASA observed microwave background no more than 3K, and concluded the entire universe of microwave temperature distribution difference less than 1K. From the perspective of heat propagation, what is the range that the 3K temperature can spread? Outside one light-year, how does the microwave background with a temperature of less than 3K sneak to the WMPA on Earth? Please let us know what this esoteric truth is that is able to fight against the basic laws of physics.

From Figure 8.5, the distance from $z = 1000$ light ball to Earth's surface is up to more than 6 billion light-years. According to the first chapter of wave propagation equation (2) for use in microwaves above the Earth measured intensity data, we can calculate the intensity of microwaves just one light-year away from the Earth. It should be $2.73k * 4\pi$ (1 light Year)^2 = 30,000,000,000,000 k. These microwaves are also subject to obey the law of from strong to weak "gradient distribution," which I discussed in "Who Should Talk about the Cosmos". From the perspective of wave propagation and the ordinary knowledge of physics, the different propagation distances away from the wave source to the observer, the microwave background from the NASA observer 1 light-year, or 10 light-years away ... how intense would the microwaves in those places would be? What is the temperature at that place?

Each wave hitting through any point P in space is a passer of point P. Is there any way to freeze them? Where does the microwave measured at P come from? They must have an emission source.

So how do microwaves flow? According to the universal law of wave flow, the propagation law of thermodynamics, or of light, heat flows from high to low in calories, and light propagation flows from greater intensity to weaker. Since the heat measured at the point Earth is 2.37k, the heat from the microwave source must increase as the distance gradually increases, while the heat will gradually decrease as the distance increases away from point P.

Unless NASA scientists can prove that the microwaves are not flowing, I cannot imagine a normal ordinary light wave in normal space that does not move. I cannot imagine microwave motion without increasing and attenuation that can spread throughout the universe! In ordinary explosions or even a nuclear explosion, the core temperature of the blast is higher than the periphery. Also, how do we deal with the interference on microwave distribution, and on measurement by huge heat energy that countless stellar and supernovae emit?

So from the perspective of light or an energy-propagation point of view, the assertion that above the Earth the microwave is isotropic uniformly distributed in whole universe is badly against the basic scientific laws.

Experts have an even more profound reason: after the Big Bang, the universe cooled down, all the big waves are gone, leaving only a little bit of 2.73k. However, after the Big Bang, did the universe continue to expand? This expansion is not uniform; it increases in distance and speed to varying degrees. Is this non-uniform spatial distribution of the expansion does not affect the heat and strength of the microwave? According to the common physical sense, we know that this is influential. Since the impact, how might the phenomenon of the whole cosmic microwave evenly distributed?

So, we have found that between the proof of the Big Bang theory supported by NASA's most powerful observations of the cosmic microwave background, and the Big Bang theory itself, there is an irreconcilable contradiction as follows: if the Big

Bang theory was established, the universe is always expanding, then the cosmic microwave background cannot be evenly distributed in this expansion space, and the so-called whole-sky map of the microwave background is a joke; if the cosmic microwave background is uniformly distributed, the microwave background sky map is correct, but there can be no expansion of space, so there was no Big Bang.

Microwave radiation intensity in the space around the light-emitting celestial

In addressing Olbers' Paradox, we emphasize that the space around a star is bright, with increasing distance from bright gradually transition to darkness.

Cosmologists seem do not care about this case. But the microwave space around these stars can be about only 3k?

Again, do we not consider the countless luminous celestial objects spreading light waves in the universe?

We have discussed a lot about this in previous chapter. Emitting objects around the microwave cannot be evenly distributed around the 2.73k, even next to the Sun. The distribution between the galaxies and star distribution in the interior galaxy space is not the same: it may be higher in a starburst galaxy, for example, and the dark space between the galaxies is probably lower.

How many human understanding inside other galaxies? Those that are considered nebula galaxies can't even be seen clearly by human being, so how can we judge whether their internal space is evenly distributed 3K microwave background?

THE INFORMATION PROMOTION OF NASA CONTRARY TO SCIENTIFIC REASONING

The size of the universe (even the NASA estimate), relative to the satellite coverage over the Earth is only one point, and is a very small point. Human detection of microwave radiation from space is limited in terms of a 'point' range. And because NASA believes that the intensity of the microwave is less than 3k, and they are all the same everywhere in the universe, is not able to expand out of reception range.

Humans have only been probing the cosmic background radiation for a few decades, which, relative to the age of the universe, is a ludicrously short time.

To extend such little information received in such a tiny space and little time segment to the whole universe is an incredible act of hubris on NASA's part.

Information **received** from a small dot inside a galaxy (note I have been using the word "receive" instead of "detection" or "measurement",) is used on behalf of the whole universe, and that is both representative of a variety of different internal galaxies, and representative of different space between galaxies. Like our analogy of the sesame seed that stands in for the whole planet, this is not scientifically sound. Therefore, as in step 5 in Figure 8.3, there is not enough convincing scientific evidence that has been gathered above the Earth to prove the Big Bang theory,

Do not at will using the Copernicus law of the universe

Look at the penultimate paragraph in bold in Figure 8.4, which describes the use of the received data above the earth to represent the entire universe because the application related with the great Copernican law of the universe: the universe, on average, looks the same from any point. Copernicus pointed out that the Earth is not in a central position. If the universe is locally isotropic, as viewed from any point, it is also uniform. So the cosmological principle states that the universe is approximately isotropic and homogeneous, as viewed by any observer.

I found some cosmological experts particularly fond of using celebrities for their own purpose. For example, the singularity inside Einstein's formula has no solution, but they felt free to take and use it. $\frac{X}{0}$ has no solution but was used in their calculations, and also corresponded it to the Big Bang, to the open model of the universe, and so forth, and said that is came from Einstein's equations. But actually Einstein didn't know any of that when he was working.

In fact, Copernicus said that at the level of large structures anywhere in the universe looks very much alike most of the rest of the universe. This is like saying the continental plates on the Earth split from each other and must therefore be able to fit together. But if in order to prove such a statement, we try to find some coastlines that match each other from across the sea, we will find that they are not a perfect match, because the crust is moving, changing due to erosion, etc. So, from a large structural point of view it is consistent, but generally not able to prove based on small details. If we think something from the Earth looks the same in all directions, can that be extended to the whole universe? From the size of a sesame seed carefully measured data can be used for the whole planet?

From the perspective of the large structures of the universe, there are also certain limits.

Previous chapters calculated that the light from our Sun can be seen up to 100 million light-years away. So, if there are intelligent beings living 2 million light years away, they will not see the Sun or know about us. So how can we communicate with other beings if we don't know each other exists?

Another example is our telescope. The telescopes we use observing at the edge of space can only see a handful of objects there. But in fact, at that observed place should be like near Earth space, same distribution of a variety of stars, galaxies and other luminous objects. Similarly, if you have aliens observing the Sun, they also cannot see the Sun, the Milky Way and other majority not particularly bright objects in the space. Well, in this case, a large structure theory obviously does not apply, because the actual existence of the universe in such a large structure is not consistent.

Great cosmologists did not seem to integrate their hypothesis. They discussed a question on the use of one relevant theory, in another case, to abandon that theory and use contradictory theories. For example, in the discussion of Olbers' Paradox, they assume a uniform distribution of celestial bodies; in time dividing the galaxy, clusters that are not evenly distributed; in the observation of distant galaxies completely forgot that in that distant place in accordance with the uniform distribution theory, there should have large and small invisible objects. Normally there are so many invisible hidden celestial objects, how dare they say the universe is 96% magic dark matter and dark energy? Another example is the extragalactic background light intensity: they count and calculate, but the latest results make people dizzy: each visible celestial body in accordance with only one photon contribution in calculation, all the photons added together to give the cosmic background brightness, several orders of magnitude bigger than the results they studied of recent decades. Cosmological experts say the universe has been expanding after the Big Bang, but also how was residual material from the explosion cooled evenly distributed? Is the microwave background? Is the distribution of the microwave at the expanding edge the same as that in the center of the explosion?

Well, here we had enough and should be able to come to a halt.

NASA has launched satellites to make images such as an infrared whole-sky map. Are infrared rays also residue from the Big Bang? What instruments can be used to measure the whole sky?

SEEING MAY BE NOT REAL - UNIVERSE DOT MATRIX TIMED IMAGE - FURTHER UNDERSTANDING OF CELESTIAL IMAGES RECEIVED

The Hubble Space Telescope has taken countless photos. People try to understand the universe by these photos, to infer the law of development of things in the universe. Many important concepts about the universe, such as black holes, are mainly derived from the images obtained from the universe. However, do these photos that we rely so heavily on really reveal the true face of the universe? Let's work together to think about it, is this a reasonable concept or not.

In order to facilitate the discussion, we need to divide the pictures of the universe obtained by human beings into several roughly categories.

Pictures of the universe can be divided according to distance into close to the Earth, and away from the solar system images. Human research space within the solar system and constellations is quite detailed. Since the distance is relatively close, these images, rough or fine, are very realistic. But this is not all. Remember in the last decade it was also reported that members of the planets of the solar system are subject to change. Even photos taken around a nearby solar system were ambiguous, so how can we talk about the edge of the universe?

For spatial image away from the solar system, we can roughly divide them into two categories: one is a very diagrammatic and one is very distant. Their measurement distance is a problem, as they are taken from the Earth's orbit around the poles of the Sun motion parallax angle to a rough estimate. The data thus obtained is a kind of rough perception of enormous cosmic objects. Another is the use of the Hubble Space Telescope and other advanced instruments to observe distant space shooting photos. These photos are very clear, and researchers have drawn a lot of conclusions from the analysis of these photographs. Furthermore, advanced telescopes have become important tools for cosmologists. People are increasingly striving to create ever more advanced telescopes. However, are advanced telescopes that capable of playing a decisive role in furthering human understanding of the distant universe?

NASA is doing so. People are thinking so. But this is completely wrong! People see are pictures of the elusive fantasy beautiful scene. People never can glimpse the real face of the distant cosmos!

For example, in the image below, assume that it is a cloud in the sky. If the distance between point A and point B is tens or hundreds of kilometers, we can be very easy at one moment to grasp the whole picture, its three-dimensional structure and hierarchy, and use the instrument to measure its size clearly.

This group of clouds is really presented in front of us, is true to the objective existence. We believe our eyes, we feel like that seeing is believing. This is our experience, this is empirical fact. However, is it the real universe?

Figure 8.6 is actually a nebula of our universe from the regiment 10 million light-years in the picture. This cosmic nebula is so big that the longitudinal distance between point A and point B is 130 million light years. Under this assumption, how would our understanding change?

In order to facilitate the observation analysis, the overall nebula is called the observation space volume G.

First, a simple fact: Suppose point A and point B of G at the same time each emitted a ray of light, the light emitted from point A needs 10 million years travel to our eyes, B point light at the same time issued will be 120 million years later than A point light come to our eyes! When the light emitted from point B reached our eyes, at the same time we see the light from point A 120 million years earlier than the light emitted by point B! That is, the same time issued lights from points A and B in G, arrive at our eyes in totally different years. The picture of G we observed is a totally time distored picture. No true information can be put together for G.

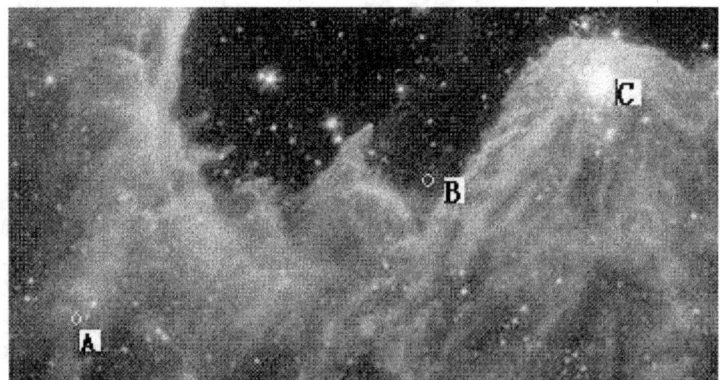

Figure 8.6 Cloud -- everyday common cloud, or cosmic nebula cloud (NASA image)

This may be somewhat difficult to understand, but what we need to remember is: Any observer cannot take the initiative to detect the stars, but must wait passively for the information arrive at the instrument. Let's do a little deeper research.

Take any three points, in G, assuming they are points A, B and C drawn in the figure.

Figure 8.7 below is a schematic view drawn from the observer's point of view. In the figure, assume plane 'g' is the observer plane, this plane may be our retina of the eye, or receive plane of the Hubble Space Telescope. Suppose the dots of light from A, B and C of G project on observing plane 'g' are a, b and c, and the distance between them are:

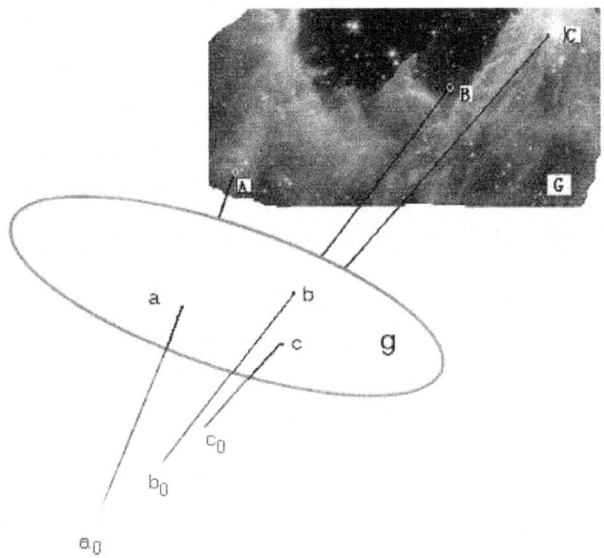

Figure 8.7 nebula observation from the observer's point of view schematic drawn.

From point A of G to point a of 'g' is 10 million light years;
From point B of G to point b of 'g' is 13 million light years;
From point C of G to point c of 'g' is 11 million light years;

Consider first that this nebula emits light toward the viewer at time T. Then the light emitted from point A at time T arrives at (T + 10) million light-years on the observer 'g' plane, at this moment lights sent at time T from point B and point C has not arrived yet. Point B light emitted at time T will reach the 'g' plane at (T + 13) million light-years later, and light from point C emitted at time T will arrive at (T + 12) million light-years to reach the 'g' plane. When the first ray of light in this nebula emit at time T arrives at plane 'g', from the observer at 'g' to observe the nebula region G, can only see this one point A.

When the light emitted from point B after traveled (T + 13) million light-years reaches the observer 'g' plane, the light hit on 'b' is the light sent from B at time T.

But the light continue sent from point A of G to 'a' of 'g' and the light emitted from A at (T + 3) million years will reach the observer. In figure 8.7, 'b' is the point

B at time T light emitted, and 'a' is the dot the light hit after emitted from point A as 3 million years later than the light emitted from B. In other words, if the B spot in G is the light from the nebula, after it arrives at 'b' to our eyes, then the light at 'a' from the point A is 3 million years older than the light from B to 'b' in the figure. We have no way of knowing if anything has changed at point A over this 3 millions of years.

The plane is shown in Figure 8.8. It is a two-dimensional plane. The three points are shown in that picture.

And this is a big area of the image that people see through a telescope, which is by the information matrix with different time mark of which reflects the difference between the different periods in the region in terms of million year units, and not at same time the region's real existence.

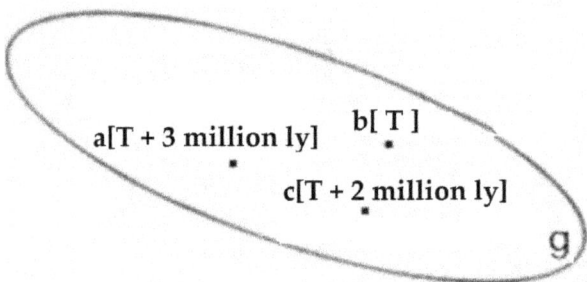

Figure 8.8 The time stamps of the lights from dots A, B and C of space G emitted at time T and arrived at observation plane 'g', projected at 'a', 'b' and 'c' on 'g'. In this observing plane 'g', the observer observed a picture composed with different dots from different ages of space G.

So in those photos from NASA, we see an aspect of millions of light-years, but in fact it is only a picture that the same as the fantasy of thousands of units of different points with different ages of unreal scene composition. Don't take them so serious like the contemporary cosmologists, counting and reasoning from them to deduce a long list stories like black holes, explosion, etc.

Seeing may not believe after all.

SEEING MAY BE NOT REAL - IS IT A REAL PLANET OR ONLY A NO SOURCE ILLUSORY LIGHT?

This is one of the many examples described in "Who Should Talk about Universe", excerpts as follows:

Looking at the star sky, countless stars greeted. For generations, people pursued

Debate of Light and Dark   Chapter 8 Dancing on a sesame seed

them and studied them. They are the main actors of the night sky, is a fascinating dream.

Have you ever thought that these stars, perhaps your favorite ones, or those you are most familiar with, may not be real stars? May just be some illusory light group?

It sounds some mysterious, but it is true truth.

We refer to Figure 8.9 to understand it.

Suppose that after the human invention of the telescope, the X star was observed at a distance from us about 2000 light-year, as shown in part 'a' in figure 8.9. In the figure, the circle represents the star, and the line indicates the distance from the star to the observer. Here we assume that it is 2000 light years.

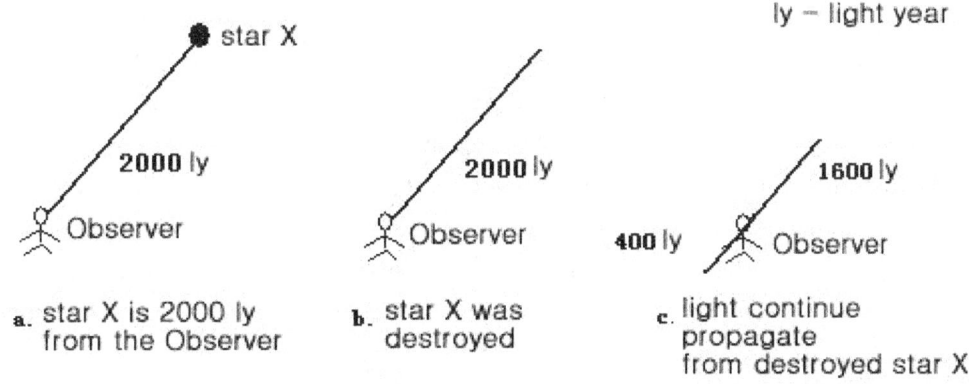

Figure 8.9 propagation of the light before and after the disappearance of the star X

Suddenly one day, the star was destroyed for some reason. But the light it has been issued is still in the transmission, and will not stop because of the destruction of the X star. The light emitted by X star in the past 2000 years to the observer continues to follow the original route and the transmission way. This is the case in part 'b' of Figure 3-2. As the star X does not exist, the corresponding representation of its circle will no longer draw out.

Assuming that from the destruction of X star to present we have observed the X star for 400 years, this situation can be used to represent the 'c' part of the figure 8.9. Well, it is frustrating that although this light had lost its source, but it is still can continue to be observed on behalf of the X star for 1600 years!

This X star has been destroyed 2000 years ago, this stars X that we have been observing for 400 years, this X star we recorded a large number of data, even actually from beginning to end for us is just a group of illusory light! X star never have existed before and after our observation! But we have been observing it! And we can continue to observe it for thousands of years!

The stars no longer exist can still be observed for many years! The distance of the X star is 2000 light years from us, and in astronomy, the distance of the 2000 light years is a small value. Most of the stars are farther away from us. Then, if those stars were destroyed, it will still exist in our field of vision for longer years.

This figure of 2000 light years is our hypothesis, in fact it can be any number, and stars at any distance, all are likely to cause such case that the star can be observed for a long time but actually never exists.

How long have mankind documented the astronomical knowledge? Thousands of years? In other words, any star on the human-recorded astronomical map, if it is more than a few thousand light years away from us, then from the recording time to the present, we never knew whether it was really there, or but had only existed long time ago, or had been destroyed during our observation.

Those stars of within the light years that the ages are smaller than human observation history, we can be sure that they exist when humans begin to observe them. This is the meaning of the distance comparison between the naked eyes and the Hubble telescope who see the farther on the time axis. The human naked eyes gather the information that can be determined by the real existence ages of the stars longer thousands of years than the young telescopes can. But are these stars that have been determined at the beginning to be destroyed in one day in recent years? We can not know. In other words, at the moment we are facing the stars of the sky, even can not be sure that any star is indeed an existing entity.

Such as the Sun, it is only 8 light-minutes away from us, that is, the light running in the space spent 8 minutes from the Sun to us. If it is at day time the Sun is extinguished at this moment, we will have to know that event after 8 minutes. We will start to feel the dark and cold because the Sun is extinguished after 8 minutes.

We see the real! But what we see does not exist is also true! The real that we see at the same time is illusory! How can we believe in our eyes?

The universe pulls up a piece of illusory distance weaving curtains in front of us. And we ourselves, seems we have not yet aware of the existence of the curtain that distorted the truth of the universe, let alone try to get rid of the illusory, to see the truth behind the curtain.

We rely on the telescope, depends on the seeing is believing, but the long year information transmission in the universe, the most unreal is the looks like to be real

The better telescope can look farther, so it may send us the more ridiculous new myth!

Looking back at NASA's thousands of stunning cosmic photos, most of them are just pages of new stories.

From the past until present, we are still standing outside the door of the mysterious truth of the universe, in the blurred illusion depicting the whole picture of the universe, and also do not see any future hope to make improvements. Facing huge universe space with slow speed that we are having, and the impossible reach distance, we are isolated from the real universe, we can not open the mystery veil of the universe! We can not see the true face of the universe!

The distortion of time and space coupled with our lack of complete understanding of what we're looking at can sometimes lead down the wrong path and bring us to false conclusions. The most important to understand the universe better, is to return to the basic spirit of science, to base on facts, to have the courage to engage in critical thinking, and also to allow and listen to different point of views.

Summary

After almost a year of drought, and finally began to rain. The leaves washed by the rain outside the window burnished, swaying in the wind and rain. Academic cosmological universe, is like a heavenly clouds blocking impossible to get the same boring shooting super imaginative science fiction film in NASA's Big Bang smoke scene.

My heart is like winter clouds as block fast enough.

Cosmological science, that which is so listless.

The true face of the universe, only Heaven knows!

CHAPTER 9

CRITICISM ON THE BIG BANG THEORY
- Century Gambling with NASA

The very premise of the Big Bang sounds so implausible. In the beginning there was void and nothingness, and then suddenly, with a huge bang, the universe and stars, heaven and earth, everything came into being.

It's hard to believe that it led to the world we live in.

I opened the window to look outside, far away mountains, near trees, blue sky black soil, Sun and Moon... how in the "bang" sound can all of these instantly appear in it? How can nothingness produce everything?

I was wondering: what were there before "bang"? Does "Universe" mean not "everything?" No universe is not without everything?

I do not like to breathe the smoke from Big Bang; I do not believe the universe was created in the big "bang" sound!

I'll have the Big Bang theory to have a say.

The Big Bang theory has a few main pillars. Let us examine and tear down them one by one..

INTRODUCING TO THE BIG BANG

I will not describe the basic concept of the Big Bang in detail here. That information is readily available from any number of reputable sources, such as Wikipedia.

There are many origin myths about the beginning of the world. In China, the myth focuses on human ancestors who opened up a new world in chaos, so we have the space that exists today. This ancient myth, in fact, is rather more reasonable than the modern Big Bang story; at least it is not so violent. It tells us that beyond the opened up universe, is still chaos. But all are ashes in the sound of the Big Bang. All these stories have been supplanted by the Big Bang. What was there before the Big Bang? What energy cause the explosion? What is outside the Big Bang universe? … All are ignored.

However, is the Big Bang real scientific theory?

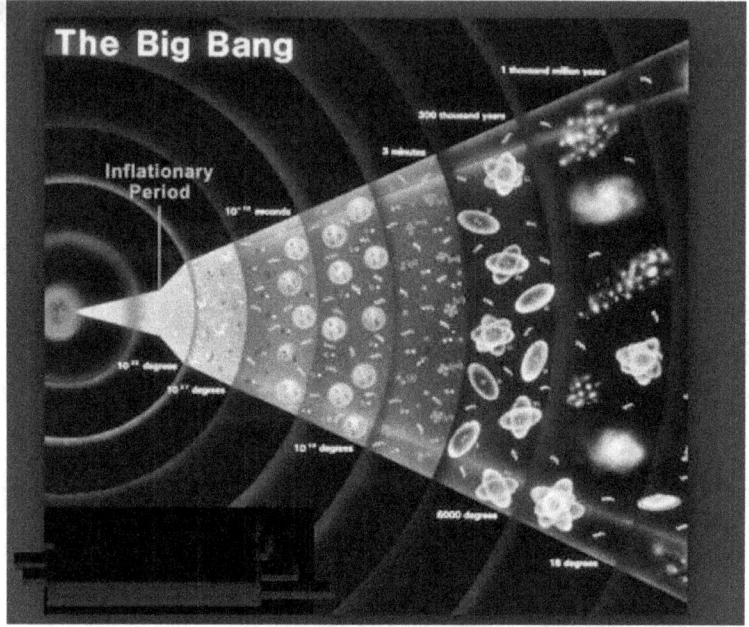

Figure 9.1 Chronology of the Big Bang

Substantive progress in the field of astronomy began from Edwin Hubble's observing distant stars through a telescope have redshift galaxies to infer the distance between the Earth and the stars. In 1929, he found that distant galaxies are moving away from Earth with their speed and the distance exactly proportional, from which observation Hubble's law is derived. To explain Hubble's law, there have been many different theories. After World War II, the expansion of the universe was explained by two opposing theories: a theory proposed by Lemaitre. George Gamow supported and improved the Big Bang theory. Gamow proposed the Big Bang nucleosynthesis theory, while his colleagues Ralph Asher Alpher and Robert Herman theoretically predicted the existence of the cosmic background radiation.

Another theory is the steady-state theory of the British astronomer Fred Hoyle and others. In the steady state cosmological model, the new substance in the galaxy away from the continuous generation of space left, so that the universe is not substantially changed at any time. History has proved that Olbers' Paradox negative steady-state theory, but with the expansion of the universe theory is given a reasonable answer, so Olbers' Paradox is regarded as one of the powerful evidence of the Big Bang theory.

Ironically, the name of the Big Bang theory of the Big Bang is from Hoyle, who opposed it. In one of the BBC broadcasts in March 1949 of "material characteristics", the Hoyle called the theory of Lemaitre "this Big Bang of view." For many years after World War II, the contest between these two theories was neck and neck.

But a series of observational evidence and other radio sources made the balance gradually tilt to the Big Bang theory. In 1965, the discovery of the cosmic background radiation cemented the plausibility of the Big Bang theory of the universe as that which best describes the origin and evolution of the universe. Now almost all cosmic physics research is published assuming the theory of the Big Bang, or is it an extension or further explanation; for example, within the framework of the Big Bang theory of how galaxies are formed, early and very early physical laws of the universe, and explanations of new observations about the Big Bang theory.

During the late 1990s and the early twenty-first century, telescope technology underwent a several major developments, and the Hubble Space Telescope expanded the horizons of space. NASA spared no expense in its design and launch and specifically the designs of the Cosmic Background Explorer, Wilkinson Microwave Anisotropy Probe, and a series of space probes to collect large amounts of data to make the Big Bang theory. Thanks to the telescope, cosmologists can more accurately measure the Big Bang model parameters, and find a lot of unexpected results, such as the fact that thee expansion of the universe is accelerating.

The Big Bang theory depends on the accuracy of many strange physical phenomena, which have not been observed in experiments, or have not been included in the standard model of particle physics. In these phenomena, dark matter is currently the most actively studied subject. Although dark matter theory still has some unresolved details, such as the galaxy halo problem and the dwarf galaxy cusp problem of cold dark matter, but it is only a matter of time before they are resolved, and they will serve to strengthen the argument for dark matter. Dark energy is another field of scientific attention, but it is still unclear whether there is any future possibility of direct observation of dark energy.

THE MAIN THEORETICAL BASIS OF THE BIG BANG

As can be seen from the above description, Hubble's law gave rise to the Big Bang theory of the expanding universe and the distance from the telescope to the redshift of distant stars redshift proportional formula. According to the common laws of explosions, naturally there will be a great deal of smoke left over after the explosion - the cosmic microwave background prophecy.

After the introduction of the theory of cosmic expansion, many cosmologists think our universe is a parallel space, and the total energy density of the universe must be equal to the critical value; at the same time, cosmologists tend to prefer the idea of a simple universe in which energy density appears in the form of material, including 4% ordinary matter and 96% dark matter and dark energy. Dark matter has not actually observed; it exists only in speculation, and has become one of the major theories supporting the Big Bang.

In many university astronomy departments, Olbers' Paradox is regarded as one of the main pieces of evidence for the Big Bang theory. Because only the expansion of the universe explains this puzzle that has plagued mankind for centuries: the night sky should be as bright as day.

Other phenomena also can be used to support the Big Bang theory, but the major helps to confirm are following supporting pillars:

- the redshift of the telescope observation that seeing distant objects, thereby laying the Hubble expansion of the universe's law;
- Olbers' Paradox.
- Dark matter theory.
- Whole sky map of the cosmic microwave background.

If these main pillars supporting the Big Bang theory are refuted, the Big Bang theory becomes untenable. All previous chapters of the book were one by one deny the Big Bang theory. The contents of the previous chapters in this chapter are a summary.

NASA'S NEGATIVE EFFECTS

The scientific community has historically been divided into two factions: those who support the Big Bang model, and those who support alternative models of the universe. Throughout the history of cosmology, the scientific community has constantly debated which cosmological model can describe the observations that best meet cosmology.

Therefore people continually revise and improve the Big Bang theory and observations to obtain better results, so that one after another give explanation of these problems.

Today's cosmologists generally favor the Big Bang model more. Support for the Big Bang theory is an overwhelming consensus; but the dissenters cannot be ignored. NASA used the Hubble Space Telescope to quell controversy – they essentially blocked the road leading to the correct answer.

After 1965, when Penzias and Robert Wilson first detected the unknown source cosmic microwave, with strong involvement of NASA, a lot of money was spent on observations of the cosmic microwave transmitter satellites, and thereby produced proof of the Big Bang sky map.

By this time, NASA had no way out: the Big Bang theory must be true; otherwise it would be a public relations nightmare. A large team of satellite launching, observation, and data processing that had been around for decades receiving funding…if it were found to be useless and wrong, it would be a disaster.

When research funding, data collection instruments, academic status is so unconditionally inclined to agree with the Big Bang theory, the formation of such non-controversial academic situation is also normal. But this is not normal for scientific research.

Present academic scholars in cosmological area, are spending their whole careers in acceptance with the Big Bang theory. We must rise up the seldom heard, but also very important dissenting voices.

Let's look at the world together to lead a NASA irrational simple example. This is not what I want to challenge the content of NASA, just a prelude to the first. But this is important because the main reason for NASA to lead the world trend is to have the cosmological data obtained from advanced tools and instruments. However, the conclusion derived from those data, is not worthy of trust, but need to make a big question mark.

Why?

First, please recall the seeing may be not real discussed in last chapter.

Then consider the following examples will know.

Ignore the simple truth, crazy world data

I mentioned this example in my book "Who should talk about cosmos". We can find many similar simple examples. I have a whole folder of them. On September 23, 2004, NASA's website has such a story (2014-12-1 still can be seen online):
http://www.nasa.gov/centers/goddard/news/topstory/2004/0831galaxymerger.html

This is a story that shocked the world, was reported in various newspapers and other media sources many of which can still be found. In this report, there are two paragraphs which demand special attention (important passages in bold):

"Here before our eyes we see the making of one of the biggest objects in the universe," said team leader Dr. Patrick Henry of the University of Hawaii. "*What was once two distinct but smaller galaxy clusters 300 million years ago is now one massive cluster in turmoil.* The AOL takeover of Time-Warner was peanuts compared to this merger," he added.

……

The observation shows the largest structures in the universe are still forming. *Abell 754 is relatively close to Earth, about 800 million light years away.*

Although at that time I had not started to seriously consider the problems about universe, after the first glance at the story my thought was: These cosmologists are wrong!

Common sense dictates that if we see any event at 800 million light years away, the event must have occurred 800 million years ago. An event that happened 300 million years ago at a distance of 800 million light-years, has a 500 million light-year-long journey to reach us.

This is a basic error that these experts should never have made.

Therefore, this report showed the daring speculation level in the name of scientific about the experts of NASA, who saw something, some of the data collected!

More than 20 well-known cosmologists, high up in the mountains, with the world's most advanced telescope, observed for more than two years, published this after brainstorming!

Then the world media vigorously followed the NASA news release, and no one objected. Of course, they are not to blame. Ordinary people, after a hundred years of science-oriented education, long ago put their trust in scientists to tell us the truth.

What scientists say is what we all believe. From this point, the contemporary cosmologists betray the trust of the people.

There is also no end to the story.

Let's look at the following piece of cut down online content from Wikipedia (2015-07-13 downloads), please note ring out with a red ellipse around the two parts of the data, is roughly translated: left red oval within the text: collision began about 300 million years ago; the right of the red oval within the text: distance 760 million light-years.

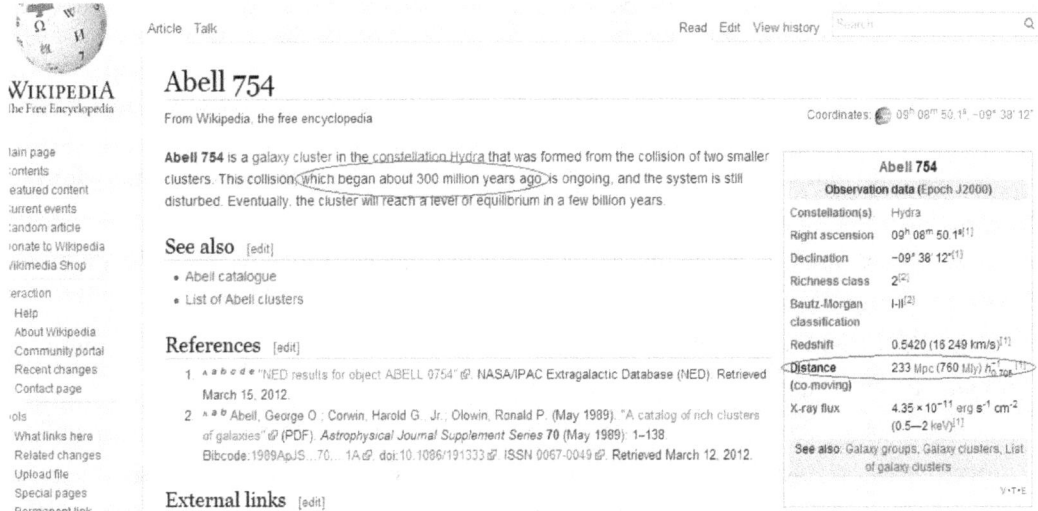

Figure 9.2 Wikipedia on Abell 754

Terrible, right? Any information about this collision 300 million years ago is still coming toward the Earth. It will need 460 million years before it reaches the Earth and is able to be observed.

Wikipedia is a site used by people all over the world, and on pages like this introduction paper about Abell754, it is assumed the readers have at least some astronomical knowledge. In other words, in more than a decade, no one of these readers in the world has criticized or corrected this error.

Today (1 December 2004) with "abell 754" keyword search in Baidu, the results that Abell754 related articles and other reports have 144,000!

I do not know in this case completely wrong script, which reported more than ten million articles, reports or related papers can produce what tricks to comedy.

The academic universe around the world, also to follow along with NASA climax to high!

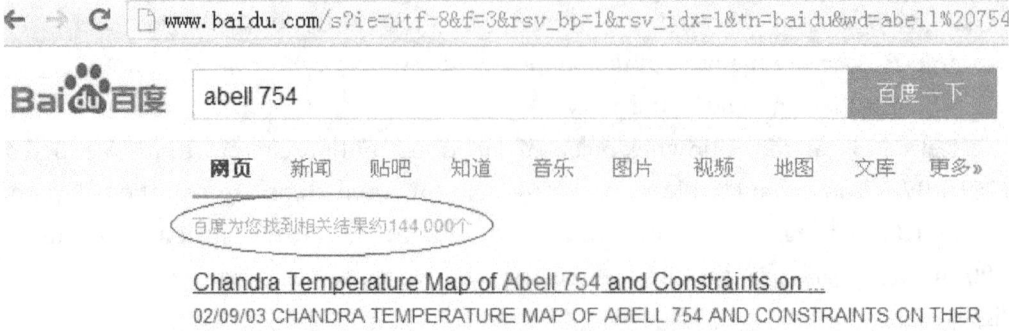

Figure 9.3 Baidu search results: 144,000 articles Results

I am the master of my view of universe – Say "no" to the Big Bang

I do not want to go against NASA, and there have been no qualifications for me to be able to be against it.

However, as long as people are considering the mysteries of the universe, they will encounter the juggernaut from NASA. There is no way around it.

Luckily my career is not in professional astronomy. Otherwise, there would be no way to stay independent and unbiased.

NASA has been promoting the Big Bang theory for half a century, spending more and more money (reportedly over about one trillion dollars in fifty years), involving more and more scientists. It is a highly prestigious organization. If a scientist can participate in a NASA project, he or she can become rich and famous, win great amounts of research funding, maybe even a Nobel Prize. If they oppose NASA's theory, their papers will tough or never be published because most famous scientists are receiving NASA project funds, so only after their approval can the paper be published. American academic journals are the world's most famous. When written papers aren't published in these magazines, their authors will lose their academic futures and never emerge in the academic community. So whenever NASA releases data, astronomers around the world will flock to use these data without thinking, even if they're obviously garbage.

If someone come up with a new theory that sounds different, and, in theory, not easily be refuted, then he or she still will be soundly knocked down. The most common way is using the newly published observational data, based on these new observing data, his/her theoretical negation from the root. That "tired light" theory in the "Tangle between Bright and Dark" chapter introduced is an example of this.

Now astronomical field is peaceful, no different voices, only the sound of the

Big Bang. From the observation instrument, the observational data to academic scholars, to journals, scientific articles, no dissent is heard.

So, NASA has dominated the cosmological world through the Big Bang theory.

In other words, regardless of how ridiculous the Big Bang theory might be, NASA has a monopoly on the market. The most typical example is Stephen Hawking, in his "Brief History of Time," wherein he wrote tartly:

"Up to now, most scientists have been too occupied with the development of new theories that describe what the universe is to ask the question why. On the other hand, the people whose business it is to ask why, the philosophers, have not been able to keep up with the advance of scientific theories. In the eighteenth century, philosophers considered the whole of human knowledge, including science, to be their field and discussed questions such as: Did the universe have a beginning? However, in the nineteenth and twentieth centuries, science became too technical and mathematical for the philosophers, or anyone else except a few specialists. Philosophers reduced the scope of their inquiries so much that Wittgenstein, the most famous philosopher of this century, said, "The sole remain task for philosophy is the analysis of language." What a comedown from the great tradition of philosophy from Aristotle to Kant!"

It is these unbridled words of Hawking that inspired me, and made me determined to write something to work against this kind of cosmologists. The result was my book, "Who Should Talk about the Cosmos,"

In fact, people who are not cosmological experts may also have a say in the issue of the universe. In today's highly developed Internet, a variety of historical and recent astronomical observation results, research results, and relevant information can be easily obtained, which makes non-professionals able to master the same information and construct different models depict a different world!

And because of the different angle of approach, perhaps there will be more convincing explanation.

NASA continues to launch new observation instrument and other space telescopes. This is actually a not normal and very unscientific approach. Following this logic, before creating a better observation instrument than the Hubble Space Telescope, those fields that were already observed by Hubble Space Telescope are no longer necessary to continue observing by other telescopes that are not that good like Hubble Space Telescope. Just take a rest. This is kind of ridiculous.

Therefore, we must try to maintain our own cosmology worldview. We do not need to lose ourselves in the fog of the Big Bang. This is not easy but it is even harder for people in academia. I sincerely hope that China's cosmologists will take the initiative and go their own way in their research, and not just blindly go along with NASA. China has its own observatories, its own scientists, own astronomical journals, and with sufficient funding, it is fully capable of opening up a new world of study and discovery.

My view of the world is my own, I am the master of my universe - I can say this because I do not need to rely on NASA for a job.

If I work for NASA, I certainly couldn't write this book!

Let us individually summarize several main points of the different views on the origin of the universe: redshift, Olbers' Paradox, dark matter theory, and the whole-sky map of the cosmic background radiation. These have been fully discussed chapter by chapter in part II, so we will mere summarize them here.

Cosmological Redshift

Sound and color frequency transforms the acoustic effect of exercise unthinkingly extended to light, resulting in a false cosmic expansion.

From redshift observation results, Doppler interprets the movement of celestial bodies as the law of cosmic expansion. Therefore, redshift is a cornerstone of the Big Bang theory.

Although we can often experience the Doppler Effect in our daily lives, using the same principles of interpretation for red shift when observing celestial motion does not comply with many exceptions. But people did not take a closer look at these anomalies, or they turned a blind eye; it does not lead in the right direction. The former reveals such a variety of huge difference between the Hubble constants obtained in different observation instruments; the latter, as an example, redshift measured from the Sun is against Hubble's Law.

We should believe the scientific literacy of cosmological scientists, when the Hubble constant observation data over many years showed differences of values, when they had been debating on their observation results, they must have been very serious did the observations, made their own conditions allow the most correct values. Then who was wrong? It can only be interpreted that the theory explaining the different observation results were wrong! Applying Doppler sound effect directly to astronomical observations of Hubble's law to interpret of the data was wrong.

Hubble's law is derived from observation, so the result is not wrong. Its mathematical form is not wrong, but its interpretation of the physical meaning of the results is wrong. The Hubble Observation constant value is not caused by the expansion movement of celestial objects, but was generated because of the celestial telescope image lag. Hubble's law was the result of manifestation, but misinterpreted the reason for why there are these values. These results will always change with different telescope precision, and these data are values of function of different telescopes, in addition with the celestial displacement effect.

A huge difference of the Hubble constant between different observation instruments, in fact, is an abnormal phenomenon, is a denial of the universe theory to explain the redshift caused by expansion.

The simplest test is to cover both your ears, cover one ear, or cover none and listen to a fast-passing train whistle. We hear the sounds in different volumes, but these different volumes have the same frequency.

Another test method is to partly shelter a live telescope, which changes the intensity of light entering the telescope, and so the redshift produced by the telescope will change. This actually occurs when the telescope uses filters. Unfortunately, people ignore this phenomenon and put it as a kind of error to deal with, and developed a set of theories to adjust.

Also, according to the theory and formulas, the observed Doppler phenomenon should appear only from far away, and in closer celestial objects that are in accordance with the Doppler formula the redshift should not be observed. But, in fact, even at close proximity of to the Sun, a small redshift can be detected. So various interpretations have emerged, including some complex equations or something like that. But neither is a satisfactory explanation.

The velocity of celestial observed image theory we discussed earlier, is to satisfactorily explain to date with all relevant redshift phenomena, and we do not need the Big Bang theory of the expanding universe as a background.

Olbers' Paradox

Olbers' Paradox is seemingly accompanied by interesting questions from people's curiosity, and has become one of the most important evidence for the Big Bang theory.

Why?

Because Olbers proved that the universe is infinite and not expanding, we should always see a bright sky everywhere, and nighttime would never come. Dark night confirms the steady state of the universe is wrong.

It appears that there were various theories direct or indirect explanations for Olbers' Paradox over time, but most of them were denied for different reasons, until there were only two theories remaining: the theory of expansion of the universe that is consistent with the Big Bang, and the theory of celestial light limited propagation time. The former is the Big Bang theory, the latter strongly supports the Big Bang theory.

So Olbers' Paradox itself became one of the main pillars of the Big Bang theory.

Wessen researched Olbers' Paradox for many years, and his theory is also evolving and modifying data, now makes the Big Bang theory completely consistent. For example, the extragalactic background light he used in 1991 was 0.7 ergs s^{-1} cm^{-2}, and in order to be consistent with the Big Bang theory in 2009, used $1.4 * 10^{-4}$ ergs s^{-1} cm^{-2}. However, it is easy to prove that this data is wrong: NASA estimated that there are about 170 billion galaxies in the **observable** universe. Even with 100 billion to calculate, each observable galaxy should send at least one photon per second to us. The average energy of a photon is $3.31 * 10^{-12}$ ergs to calculate the extragalactic background light should be between 33.1 ergs s^{-1} cm^{-2}, significantly more than the 2009 data used by Wessen, but is close to what he used in 1991 value.

Of course, the extragalactic background light data he used did not came up with entirely on his own, but the results of observational studies that were actually the product of NASA's advanced telescope. But this result after nearly 20 years of hard work has obviously let him farther and farther down the wrong path, and I have to say it is a great irony for NASA's massive funds to be used for a corroboration to that!

The layer of the shell model used in Olbers' paradox itself is not correct. The distribution of luminous objects in the local space is uneven. Although the theory of the universe from the big structural point of view is not so uneven, when it comes to a few light years, within a very small space of dozens of light-years, the luminous objects are often very unevenly distributed. Otherwise, why do we give celestial combination names of stars, galaxies, clusters? Because the names are based on the distance between the celestial bodies, the small space distribution of luminous celestial bodies plays an important role in this why-is-space-dark problem.

Olbers' Paradox, as well as numerous astronomical models, like the microwave background intensity, extragalactic background light intensity, etc., do not consider the distribution of luminous objects in the universe, how they affect near or distant space light intensity. Like the partial local distribution of celestial bodies such as "two suns", has a deadly effect on the conclusion, was also not taken into account in Olbers' paradox solution history.

Based on the specific mathematical models in chapter one, "A Tangle between Bright and Dark," it is proved that Olbers' Paradox does not hold. Thus, the Big Bang theory has lost another pillar.

Whole-sky map of the cosmic microwave background

The whole-sky map logically is not established; we have a whole list that shows the error of their logic. With vast amounts of money to do long-term logic errors, a large number of observations, the results are not likely to be correct.

Many cosmologists will always unconsciously use their telescopes to look directly at celestial bodies. However, the thing you see is outdated and old, then you see the light wave's rapid attenuation with distance, and see an unreal composite image of different ages. When the lights from different distances reach the observation point, their intensities are completely different. If we had a uniform distribution of microwave measurements surrounding the Earth, then following the light or heat formula of the law of propagation, the various microwave distribution left to different distances from the Earth would be completely different. Even values measured at the Earth surrounding proved that hugely different values would be at different spatial distances.

Applying "because I cannot find (), it must be ()" does not sound logical. We have shown that there are three kinds of hidden celestial objects that are beyond our vision, but it is entirely possible to play the source role of the measured microwave background. We must not abandon them in the forgotten corners just because we lost sight of them, they are likely to be the source of the microwave background.

The concept of the law of large structures of the universe did not undergo a rigorous proof. It is the product of a logical thinking. There are still specific conditions to the application of large structure theory. Especially as the super structure of the universe compared to the Earth is a relatively small entity, it gives a ridiculous feeling. In the case had been endorsed by famous people, but actually basically has not and impossible can do the strict proof on the theory, and based on it to carry out large-scale satellite observations for decades, which is not a very smart thing.

When the test model itself has no scientific base, then accurate data calculated and mysterious model used, the joint produced is only "garbage in garbage out."

Seeing may not be real. Burst the cosmic psychedelic image NASA issued with its most advanced telescopes. Ask them please not to take so much confidence with these illusory pictures.

The major substance of the universe is hidden celestial objects, not dark matter

Dark matter is said to be the product of the cooling process after the Big Bang. 96% of the substances in the universe is dark matter. And this magic material that is not bright and does not emit nor absorb light substance, has never really been discovered. All claims to the discovery of dark matter have been proved to be hidden objects.

I also think there is more than 96% of invisible material in the universe, but there is no scientific basis to say that it is dark matter.

From the composition of matter, from the known atom, the known molecule, how can dark matter neither absorb light nor emit or reflect light but still have a gravitational field?. And for there to be such a huge amount of large-scale dark matter that human for a long time actually did not understand it belittles human intelligence.

According to experts, the cosmic microwave background proved with the magical cosmic large structure of the law as an example, during all of human history, including aircraft, Mars exploration, microwave, etc. No satellite spacecraft has ever collided with dark matter. Then we have more proof that microwave background exists, and the shape of the dark matter similar real entities do not exist!

With the normal way of thinking to consider the problem of dark matter and the hidden objects:

1. At the farthest distance the telescope can see, the telescope can only see a small number of superstars and other particularly large objects. But in that place, according to the law of large structures of the universe, the distribution of celestial bodies in general should be distributed almost the same as the space surrounding our galaxy. That is, in that space are also distributed many clusters, galaxies, stars and planets, but we do not see them. These invisible objects are hidden objects, so how many of them make up 96% of the so-called invisible dark matter and dark energy?

2. All the discovery of dark matter hinged on the effect of the gravitational field, without exception, were later proved to be ordinary celestial objects. But at the time

they were found, the telescopes were not sufficiently advanced, so the celestial bodies were hidden to our eyes. When more advanced telescopes were developed, temporary hidden celestial objects turned into observable celestial objects.

3. To determine the sheer number of hidden objects, we know that by comparing the results down to lift the telescope and the different observing results. Those who put down the telescope cannot see, but with the telescope are able to see celestial objects hidden from the eyes. With the exception of the Large and Small Magellanic Clouds, all the rest of the galaxies are hidden to the eyes!

4. Similarly, when telescopes became more advanced, they found more and more visible celestial objects. Therefore, the behavior of humans making more advanced telescopes proves that there are many hidden celestial objects that have yet not been found.

These phenomena are obvious. Taking them together, what percentage of the dark matter have they occupied? Remove these hidden celestial bodies from the 96% and what percentage remains? Why did the cosmological experts turn a blind eye, and not want to think about, and not be willing to challenge the theory of dark matter?

Even if the "hair" in the form of dark matter exists around the Earth, we also need to prove whether it exists in the vicinity of the stars like Sun, then it will be possible to estimate the amount.

The fundamental problem is that dark matter is from the basic structure of the universe composed of matter formed, to justify the Big Bang theory. Once dark matter does not exist or does but only in a very small amount, then the basic structure of the universe with the Big Bang theory formation process is finished. After decades of thousands of researchers, funding, etc. efforts, forming an overwhelming advantage, the Big Bang theory has won out over all other theories, and several generations of researchers of the cosmology in the Big Bang are growing up secure in the knowledge that it is correct, so how could dark matter be recognized it is non-existent?

Recall among Figure 2.10 the universal distribution map of the hidden celestial objects. We know the number of visible objects, even by IC1101 absolute limited observable radius to hypothesis set the size of the universe, only accounts for a very small very small part for the hidden celestial object world. The universe known to humans (13.7 billion light years to calculate) is less than 0.003% IC1101 delineated universe size.

The universe, with a distance limited us in such a small cage.

FOLLOW NASA to consume Copernicus' Law of Universe

Here, I also want to follow the cosmology experts' consumer laws of the universe of Copernicus, to use it to prove the existence of dark matter, or to prove that 95 percent of the invisible substance is in fact, hidden celestial objects.

The proof is as follows: the laws of the universe are derived from Copernicus's theory that the Earth is not the center of the solar system or the universe: to the observer standing on the Earth from the big structure, large-scale point of view, the universe is close to isotropic, homogeneous distribution. So the cosmic background radiation should look substantially isotropic. According to the results of microwave background space probes in the vicinity of the Earth over decades of receiving data, we have drawn the whole-sky map of microwave background. Now according to the same logic that cosmologists using Copernicus's law, we can determine no presence of dark matter in the universe!

Because ninety-six percent of celestial objects are said to be dark matter. So no matter what they are, they are uniformly distributed in the universe, so we infer that space near the Earth should also have a uniform distribution of dark matter. And because dark matter accounts for 96% of the matter in the universe, it cannot be further than the thin air of the substances. If spacecraft were to encounter it, it may be perceived.

But we have fired many satellites, rockets, spacecraft, and other space probes (COBE, WMPI other microwave background detectors also just contained therein limited number) at different heights above the Earth for many years, and these aircraft have not reported any hit or found any dark matter. Therefore, to detect dark matter in the vicinity of the Earth, we did much more exploration than COBE or WMPI did for microwave background, more comprehensive more thorough coverage much larger of the search space, much longer observation time. And our exploration for dark matter also included COBE, etc.

So, if we follow the "part to whole" same magic logic used by NASA to infer the whole-sky map, that "data collected from the local universe can be extended to the whole sky", we can infer that in the universe, there is little dark matter because a large number of Earth probes did not find any dark matter, and this dark matter probe covered much more space than the whole-sky map probe space. Thus if we can apply the law of the universe according to NASA experts' logic, can we determine that there is no existence of dark matter in the universe?

Dark matter does not exist; what we think is dark matter is simply hidden celestial objects.

A Century Bet with NASA

NASA experts are trying to find cosmic dark matter. They consider finding dark matter to be the next stage of their most important task. However, according to various reasons already described, I do not believe they can succeed!

I am here make a 100-year gambling promise with NASA cosmology experts:

I wager that in one hundred years, NASA will not find more dark matter than hidden celestial objects! I say that it is impossible to find 30% of dark matter in the universe!

I was going to say, give a hundred years, you could not find a few (I'm not entirely sure exactly dark matter does not exist, or can't artificially make a few) dark matter, not to say find out such huge dark matter like a planet! However, for insurance purposes, from a scientific point of view, better to compare to hidden celestial objects. Remember, as we calculated, the Sun is only visible at a distance of about a million light-years. So we can never see those suns that exist a million light years away or farther, let alone stars that are weaker than the Sun! Based on the theory of cosmic large structure (here is an appropriate place to use the theory), we can only see a small star at the edge of the universe, in fact, as we see the Milky Way near the same space, there are the distribution of many large and small size clusters, galaxies, stars and other celestial bodies. We define these objects as hidden celestial objects. Since there are so many hidden celestial objects, so if we subtract these hidden celestial objects, what is the rest of the percentage of dark matter? If dark matter exists.

If the universe does not have sufficient quantity of dark matter, the material basis of the Big Bang theory has lost its fundamental support.

Since I have been writing this book for more than a decade, the bet began a decade ago. But I will give some extra time to have this bet start from the official publication of the book, which should be in 2016. Let's see the outcome in 2116! One hundred years later to see the outcome!

Although perhaps I would not personally see the result of years later, but in my lifetime, I will always enjoy my victory!

Using concept of hidden celestial objects to test the concept of the expansion of the universe

We can use the concept of hidden objects to verify the correct theory of expansion of the universe.

In accordance with the discussion of hidden celestial objects, according to the sixth class of temporarily hidden celestial object (distance, increase), if the theory of the expansion of the universe is correct, then since the distance continues to increase, many among hidden and non-hidden celestial, after the expansion of the universe, their distances from the Earth increased, so they should from observable celestial turn into hidden celestial objects.

In "Who Should Talk about Cosmos" I compared the difference of observing the universe between with human eye, and with using a telescope, and pointed out the human eye is not nothing. In this chapter, we will appreciate the advantages of the human eye to see specific examples of the universe.

We can make a comparison of various ancient astronomical star maps to modern maps observed with the naked eyes, and check if the stars appear in the ancient astronomical maps have disappeared or not in today's visual observation, If no such objects disappear, it means the expanding universe theory is doubtful.

We can also use the same antique telescopes and binoculars used hundreds of years ago, under the same conditions, to observe the stars of today, and compare the results obtained to see if there any of the celestial bodies observed long ago have disappeared. Because of the expansion of the universe, even if now the observed objects do not disappear, they must also blur than previously observed, and the brightness is reduced. Under the same conditions as much as possible the results of comparison over time, we can see whether there is the impact of universe expansion.

If you can see some ancient, barely-observed objects that have disappeared today, it shows expansion of the universe is real. Even if only a handful of celestial bodies, or none, are still visible, it means expanding universe theory also questionable necessary.

We can use the records of any historic human observer, who described distant objects, and then we can use the same type of observation instruments to compare past and present images obtained under the same conditions of observation, because the expansion of the universe, now observed pattern does not disappear then, certainly the blur was worse than previously observed, brightness is reduced. Different period results comparing, we can see whether there was the impact of expansion of the universe. For example, the earliest recorded observations of the Andromeda galaxy may have come from the Persian astronomer Arz, his book describes the Andromeda galaxy as a "small cloud". In 1612 German astronomer Simon Marius first observed and recorded the Andromeda galaxy with a telescope. It shaped like candlelight seen in a horn tube.

We can use the same instruments as our predecessors to observe the Andromeda galaxy and compare the results of two observed images. If the universe is expanding, our image of the Andromeda galaxy, which is about 250 light-years away from Earth, will be affected to some degree.

The foreseeable future, human beings cannot know the true face of the universe

Where did the Big Bang start?
Before the Big Bang, what did the universe look like?
What exploded?
After the explosion, what was outside the explosion space?
We do not know the answers to these questions.

That being the case, why don't experts take a small step back off the Big Bang itself, so they do not categorically pursue good? The absolute limited observable radius of the universe defining human use no matter how advanced telescopes obtained vision, cannot piece through the absolute limited observable radius of the universe. And mankind has only mastered very slow speed of space travel, but the human footprint in the universe can only reach a very small range in a fairly long historical period within the solar system, and could not fly out of the range of about one light year.

The universe covers up its true face with veil of distance. We cannot travel fast enough to push away its veil. The primary task of humankind, is not to think of ourselves as closer to understanding the truth of the universe, but to recognize our enormous limitations, to strive to improve our speed, try to understand that our surrounding space can be appreciated, and put our efforts into finding planets we might be able to settle on, something that would be beneficial to our fellow humans.

Using up our whole short century of life,
explore the endless mysteries of the universe;
Consumption of human resource accumulation in the millennium, to find the essence of more up and down the horizon maze.

In short life seek eternal fortune,
holding turtle speed and reluctantly climbing the ladder to heaven.
Wake up, wake up,
Long star way to go,
    for us to carefully understand and search;
Far mind span to cross,
    for us to be sensible tracing.
Using the smart thought of the human intelligence,
    drive the ark to get the obtainable truth!

PART III
CALLS FOR THE CREATION

Mankind is losing the ability of theory innovation, mathematics is getting old.

- Calls for the creation
    How to stimulate innovation capability?
    Modern education stifles the creation
    Cost of creation
    Call for creation,
- Mathematics is old
    Why math is getting old? – Climbing trees or planting a tree
    "End of Science" comparing to the rise of science creative opportunities.
    How to make science rejuvenated?

> Wise man must be a loss;
> Fool consider one thousand, there must be gains.
> --- "Records of the Historian Huaiyin Hou Biography"

When growing calls for freedom and breakthrough,
With a transparent or multicolored ribbon for him we draw up the plan holder bin;
When the wild wind is going to blow apart the creative pods,
But a transparent beautiful greenhouse shielding away the harvest wind;
Impulsive thinking like a kite pursuing free arms of clouds in the blue sky,
The secular link ties always make it stumble, disorient;

Professors in the beautiful resort smugly command the denounced deviant crackpot,

Experts carrying creative banner seeking a world that inside may accommodate of their aesthetics think.

We call creation,

Creation, but do not know where it is wandering.

Thoughts like these are often considered part of the problem for me: when the development of human science and technology is taken to the extreme, what kind of impact on the process of the human mastery of knowledge will be produced?

First, after resolving various functions in the human brain, the human memory will be able to selectively and/or complete copy down the data like copying it to a computer hard drive;

Then, it will be possible to copy down the various categories of knowledge, collating and storing them in the repository knowledge data base;

Finally, it will be possible to put the knowledge data in the repository to be selectively copied into the human brain.

If the above scenario occurs, the human learning process will eventually be a simple procedure that is not painful. When somebody reaches a certain age, he can go to choose, and within a few minutes, copy selected knowledge into his brain.

However, this knowledge that can be copied and stored is only static data, not creative knowledge.

We may be able to replicate knowledge in the future, but not creation.

To human progress, creation is more important power than knowledge.

This chapter has some controversial and even outrageous arguments and topics. In fact, the whole book does.

We propose to discuss the problem, hoping to play a valuable role, arousing attention and discussion on issues of scientific progress.

If you can grasp the opportunity to emancipate the mind, clarify your ideas, and find the right way forward, you might be able to promote China's science and help it leap into the next room for development. So, we want to try to trace the hit new lows, analysis to explore all possible factors that hinder creation.

Here, that creation occurs in such inventive acts like an "algebraic analysis of complex systems."

CHAPTER 10

FOSTER THE CREATION

THE PURPOSE OF EDUCATION IN MODERN SOCIETY

After the Guardian appears, that is, after the success of the Industrial Revolution, the purpose of human education is to cultivate a qualified workforce and train qualified engineers and technicians for the industrial production line.

Until today, this purpose has not changed, but only intensified.

We easily found on the Internet some of the definitions of "educational purpose." Here is a brief summary:

- The purpose of education is to train people to become useful to society.
- Though there are a lot of ways to learn, but centralized education (especially schools) has been proven by history to be a better way.
- To become educated, people need to be trained to meet general requirements of certain social needs.
- In general, educational purposes often sketch out some ideal society under the guidance of philosophy to indicate the direction of education efforts for its citizens.
- Educational purposes have certain ideals. The origin is not the education itself, but rather social, cultural, and philosophical views that influence educational decisions.
- After an abstract and general "philosophy" of things,
- Special education goals: the goal of the country at this stage of education is "for the economic construction."
- Educational objectives are clearer and more specific than educational purposes.
- The ultimate goal of education is to train students to become useful to society and to meet general requirements for certain social needs.

In modern society, school education is the only way. Although self-study is possible at the elementary and secondary school level, no individual can own the educational resources that an outstanding university has. To be at the forefront of research in the field, entering the university education system is the only way.

From education purpose to implement educational, we can understand a few things:

Modern education is training a useful person to society; its goal is train people to be able to make some contribution to society. Because modern society tries to promote equality, basically "everyone is equal in terms of education," and while schools in countries have a large difference in the quality of education, but the overall educational goals and teaching content is not essentially different. Thus, education policy-making can only meet the primary needs of the community, and the primary social demands are for a qualified, talented workforce. Therefore, a primary goal of education is to cultivate a qualified workforce.

That is, education after the Industrial Revolution, from primary school to university, is substantially the same in the pursuit of the goal to meet the main needs of the community for the purpose of providing a wide variety of talents for the society. In this globally consistent goal, all basic schools are marked with the stigma of mechanized production and the output of standard components that can be used in the big machine of the community.

Until the graduate study stage, students are not out of the study and exam pattern.

This mechanized production model has produced generations of standard output, although each scholar, professor, or researcher does their utmost to create something new. But when the ideas of different individuals have been unconsciously unified by years of school education and shaped into standard form, where is the inspiration and courage to break out and do something revolutionary? When all the trees have been toughened, who may plant a new tree to live in the steel environment?

The age after the Industrial Revolution is characterized by a pursuit of high-efficiency. And in this era of mass production, technology development is one of the features of the second phase of the Industrial Revolution.

The two main methods for large scale production were developed in the United States:

One way is to create standard, interchangeable parts, which then requires a minimum amount of manual labor to assemble these parts into a complete unit;

The second method is to design the assembly line.

The method that the schools use to produce a large number of graduates and the method for large-scale industrial production are much alike! In a globalized world, no matter from what corner of the world, from a dusty town in Kenya to romantic Paris, the school will quickly erase personal characteristics and arm each student's mind with the same way of thinking, the true realization of a standardization of the way people think.

When all the people use the forest guardian's way of thinking – which can be called scientific method or some term approach – is when all human beings will invariably be caught in the vicious circle of lack of creativity.

Before the Industrial Revolution, human education, which for thousands of years there have been a wide variety of trees planted down during that period, was not so mechanized or subject to standards. The purpose of education was not explicitly for profit services, thus promising a variety of seemingly absurd arguments were more tolerated. Strange ideas and the freedom to discuss and research them, can make a variety of possible germs emerge. At that time, the ground was not been so hardened with a cement covering, and it was easier to break, right?

MODERN EDUCATION STIFLES THE CREATION

When I was a graduate student at China Mining and Technology, I had an incredible experience.

In the early 1980s, the political class for graduate students in China was "Nature Dialectics," part in which introduced some of the latest theories about the universe model. I found it very interesting, but after thinking about it, I got some different ideas, so I spoke out and discussed them in the classroom with the teacher. Since I was talking very heatedly about the content of the textbook, I reckon that this teacher, known for his rigorous scholarship, felt very unsatisfied. Due to the relatively lively discussion, the content of which was against the textbook, and my doubt about Einstein's relativity theory, he was even more unsatisfied. He told me that if I did not get the right ideas and submit a deep review, I would not pass! I really had to go against my nature to write a review report, which got 75 points, the lowest score I ever got for a lesson.

Figure 10.1 Knowledge, ah, only you can lead us through the night

This approach leads to the direct consequence that the individual cannot have unconventional thoughts and is stuck in a rut. Is this Natural Dialectics?

A friend of ours told us about his own high-school experience in Hong Kong.

He was thinking about how theoretical interference leads to uncertainty so that the instrument cannot accurately measure particles. So, for photons, which are even subtler than atoms, there should be even more uncertainty. Consequently, his thought was that the so-called two-phase wave-particle nature of light may be only a consequence of the uncertainty and not the true nature of light.

After the teacher listened to his argument, he was prohibited from continuing to think so, because if he did, he might not pass the test for university study!

He never went back to thinking about such an abstract problem.

Until he read *"Who Should Talk about Cosmos,"* which brought back his old memories.

I said, what a shame, but if I were your teacher, you have to let repented, or don't mention getting into the Tsinghua University, even the second universities being good trouble. You might have hoped to get the Nobel Prize, but it's too far away. To sneak into the revolutionary ranks first is more important. Otherwise, how can you continue revolution?

Whether it is in order to maintain their own authority, or for the sake of the future of the students, industrialized education, exam-targeted education, does not fit with whimsicality, deviant innovation, and creativity.

Set up the learning base for creation

Since the primary purpose of the establishment of a national education system is to serve the needs of society as a whole, it is therefore impossible that it satisfy everyone's needs. While most people's learning objectives are consistent with national educational purposes, there are some people who need to set up their own special learning objectives.

In fact, the enrollment of students in special schools for arts or sports is an example of the establishment of special learning purposes, special learning environments, and special teaching methods.

So, in order to cultivate creative talents, training talents that may change the status of the theory of human science, should our country recruit students with special creative talent and provide an ideal base for those students, a growth and learning education environment that is unlike modern standards?

The idea is to approve the selection of students, establish some traditional breakthrough standards, and create a free and relaxed environment allowing full freedom of intense academic debate. The purpose of their study is theory innovation, and the birth of a new theory, in which they are expected to be able to make a breakthrough contribution to mankind.

This may take one or more generations to accomplish. But the investment is not wasted. Even if they have not planted new trees, they can master the fields of science and technology theory, and success at will is certain.

There are many factors involved in all aspects of the need to break the traditional thinking to do eclectic guide. It sounds like another utopia, and there is no need to continue the YY. (YY is a network term, an "obscenity" acronym with meaning similar to our fantasies here.)

However, this prospect is very attractive. We have hope that we can cultivate excellent talent for scientific and technological innovation theory without huge costs.

THE COST OF CREATION

The cost of creation is enormous.

This is also deviating away from the efficient and economic principles of modern society, especially those projects that cannot see the practical value of study and that almost no one is willing to invest in.

One example is Boolean algebra. After being invented, it took almost a century before it found its application in the fields of automation and computers. If Boole had to apply for research funding to carry out his Boolean algebra project, nobody would have given him a penny and he might not even have invented it successfully, nor would the authorities that oversee journals publish his work.

In modern society, the cost of creation is too high. And because of the sharing and flow of information, the wealth of creation can easily be stolen by someone else, even faster than the creator himself can capitalize on the creative achievements.

The risk of creation is also high. A person may study for a lifetime but get no results.

Application research professors can engage with their colleagues in pure research as "play": they play with money, play with time, and do things that may never turn into marketable products. This is incompatible with the basic principles of modern society and the social value and orientation of mainstream science investment, and it is difficult to get enough financial support.

Thus, if a modern man can stick to doing research in the field of pure theory, he will quickly find it extremely difficult or frustrating to stick out.

And the people who keep the faith have to endure poverty and loneliness. Insisting on the fruitlessness of confusion and facing the pain seems no obvious hope. He may, oh no, in the eyes of the laity, be seen as a Frankenstein.

So, now most professors become fundraisers, going everywhere to get money, and then they use the money to bring in assistants and graduate students, who buckle down to specific activities so that they can bring in more funds.

Nowadays, the qualification for professor competitions is to see who has more high-quality papers. How many papers can a person put out in a year even writing to the point of exhaustion? Therefore, the number of papers published is actually based on who gets more research funding: more funding means one can recruit more students each year and thus can publish more papers, which leads to a greater fame in various fields and thus more money. Therefore, a famous professor can work here as a senior consultant and there under the auspices of the International Conference, as well as publish more academic papers than others. Of course, many times he would be the first author in general, followed by the student who actually did the work.

Last weekend, I was invited to a performance at the University of California, San Francisco. It was a Chinese New Year celebration for the show, and I learned that in this world's top medical schools, there are more than 800 leading scholars from China. Professors pay them below market price (called a post-doctoral) to work here and write articles.

So those who engage in pure research also would like to be closer to the applications. They do not want to stay in the underfunded areas that usually make up theoretical work.

In such an economy driven by the trend, the power of innovation for scientific theories is very faint, ah.

But if there is no theory innovation, Horgan saying that science is ended is a great truth!

To foster creativity, we have to get rid of the unified mechanization mode of thinking

In view of the fact that training creators with the mechanized assembly line of education is a complete failure, culture creators first need to abandon the unified mechanized mass production of casting molds in order to nurture a new way of thinking and organization.

First, we need to get rid of the unified school structure and mode of education stand from the steel belt to earth ground.

There is a style of teaching for Chinese literature classes in middle school that was once known as "Knowledge Tree":

Six textbooks have 180 lessons and more than two hundred articles (including poems). The main intent of the editor is that students don't just read an article but, more importantly, through the study of materials, master the system of language and improve reading and writing skills.

The first level of the six books consists roughly of four parts: the basics, classical, literary knowledge, and reading and writing.

The second level of knowledge divides the first level into the following 23 areas:

The basics includes eight areas: voice, text, words, sentences, grammar, rhetoric, logic, punctuation.

Classical includes four subjects: word, form words, function words, sentences.

Text knowledge include four areas: foreign, ancient, modern, and contemporary.

Reading and writing includes six: centers, material selection, structure, expression, language, and genre.

The third level of knowledge consists of the 23 aspects of the second level applied to more than 130 knowledge points such as syntax, parts of speech, phrases (now called phrase), single sentences, and complex sentences.

Using analogy, the structure of this language knowledge is presented as a diagram of traffic patterns in China: the first level of knowledge is represented by the provinces, the second level of knowledge by the cities, the third level of knowledge by the county, and the more detailed knowledge of the third level by towns and villages.

......

This junior-high language knowledge tree is exactly in accordance with the organization of the scientific method.

We see this tree of knowledge and cannot help but have a feeling of horror, seeing the flesh-and-blood body dissected and scraped by the bone pickers as they categorize.

We dare to break down the language: this method for breaking down an article can teach a student to become a good critic, but not to write any good articles or literature. Because the quality of an article is not merely its literal meaning but also its subtext, it's meaning between the lines. Article style, taste, and even genre and language characteristics cannot be dissected. The meaning behind the text, such as satire, praise, etc., cannot be easily split out with a scalpel. If a student, skilled in the application of the scientific method, uses this standard classification system when he reads every article, every literary work, then, for him, literature is a dried out string of bacon or sausage, hung up by category, and with a musty smell.

In real life, how many writers are trained out from the Department of Chinese in university? The scientific method can train a critic, but it doesn't provide what is required to train a creative writer.

In other words, the method of scientific analysis is extremely unfavorable for the appreciation and creation of literary works!

Students educated in this method of scientific analysis are good at being critics, technical writers, historians, secretaries, copywriters ... but not at being writers whose main work is creation.

Writers can never be cultivated by teaching, not with the knowledge of scientific anatomy to develop.

Writers in the Republic of China: Shen, Congwen and Zhang, Henshui never graduated from high school; Hu, Shi studied philosophy; Lu, Xun and Guo, Moruo learned medicine; Xu, Zhimo and Yu, Dafu held a degree in economics; and Zhou, Zuoren first went to naval school, then studied Greek.

They have not studied literature, and there was no one to teach them writing. Talent and self-study made them become great literary people. The method of scientific analysis is not free to be introduced to any field!

But the method of scientific analysis is dominating the world's way of thinking!

The method of scientific analysis focuses on the details of the categories while ignoring the overall picture, paying attention to minor details while ignoring the torso, and makes it difficult to foster excitement and a desire for innovation. It is difficult to think of planting new trees with this way of thinking!

Perhaps this—the fact that scientific analysis has been the dominant ideology—is why we still haven't seen a new tree planted for many long years!

Should we have some introspection about our way of thinking?

INSPIRATION OBTAINED FROM THE INVENTION OF BOOLEAN ALGEBRA

In 1835, 20-year-old George Boole opened a private school. In order to give students the necessary mathematical courses, he read with interest some introductory mathematics books. Soon, he was surprised: this was math? Unbelievable.

So, the youth who'd received only an elementary mathematics education self-studied the abstruse "Celestial Mechanics" and abstract "Analytical Mechanics." Because he had a strong feeling of the beauty of symmetry in algebraic relationships, in his lonely studies, he was the first one to discover the invariants, and a paper about the results was published.

After this high-quality paper was published, Boole remained a school teacher and started communications and exchanges with many of the foremost British mathematicians. Among them was the mathematician and logician De Morgan. De Morgan, in the first half of the 19th century, was involved in a famous controversy. Boole thought De Morgan was right and published a slim pamphlet in 1848 to support De Morgan's argument. This book became one of his greater achievements six years later. When it came out, it was immediately praised by De Morgan, who also congratulated him on opening up a new and difficult research field.

At this time, Boole had been studying algebra logic, which became known as Boolean algebra. He simplified logic into a very easy and simple algebra. In this algebra, "reasoning" is expressed in arithmetic formulas, and these formulas are much simpler than the ones that for most of the past the second year high-school algebra course used. Thus, the logic itself accepted the domination of mathematics.

In order to make his work perfect, over the next six years, Boole paid an unusual effort to his work. In 1854, he published *The Laws of Thought*, his masterpiece. He was 39 years old. Boolean algebra was born, and the history of mathematics erected a new milestone, its youngest tree growing in the forest of mathematics!

Like almost all emerging knowledge, no attention was paid after the invention of Boolean algebra. Continental Europe's leading mathematicians contemptuously called it mathematically meaningless and philosophically strange. Based on this, they never suspected that this mathematician from the British island was able to make a unique contribution to mathematics.

Soon after his masterpiece was published, Boole died.

Early in the 1900s, Russell's *Principia Mathematica* said that pure mathematics was found by Boole in his *Laws of Thought*. Once this came out, the world's attention on Boolean algebra was immediately aroused.

Today, Boolean algebra has been developed into a major branch of pure mathematics.

In recent decades, Boolean algebra has had a decisive importance in the current application of the most important engineering and technical fields of automation technology, computer logic design, and the like.

From this tree, which is basically regarded as the last one of the few trees planted in the human mathematical forest, the invention experience of Boolean algebra can be seen:

Boole did not accept a unified learning and training in the form of modern universities, so his thought was not portrayed as a standard model of industrialization;

The friendship of Boole with De Morgan, and his participation in academic debate, triggered his inspiration;

Boole's work was not recognized by the authority in his day, and he was even ridiculed by them. In other words, Boole's papers could not be published at the time by the leading academic journals.

Boole's research, over a long period of time and about a century later, was the foundation of computer science and, as a theory, has been reused in the field of applied sciences.

Ideal for educational purposes

In fact, we discuss this topic is like an empty talk a little over our position.

To discuss the ideal educational purposes, we have to talk about the ideal society. As they are already intertwined, the educational needs of the students need to be consistent with social form. Otherwise, the educational process will create social conflicts, making the "ideal" word irrelevant.

Further, as mentioned earlier, if the development of science and technology through the mind of the administrator of the computer knowledge, then all forms of education become meaningless.

When knowledge can be selectively copied into the brain, the only requirement for learning is innovation.

We would like to discuss a little bit the question of, in the current form of society, what kind of ideal aims of education might be established?

Students can receive an equal education for free;

Students can learn according to their own ability and efforts and have equitable access to different higher education institutions;

Students can graduate with fair employment;

Companies can recruit qualified personnel to their needs;

Institutes should be able to recruit enough talent;

Government should be able to obtain qualified managers;

... ..

People are realistic, the subject of getting on innovative creativity and talent cultivation will certainly after meet the basic living conditions, etc. Food enough then with knowledge etiquette, since ancient times is the case. It is the same talking about idealized education.

We are here to open a small beginning; in-depth discussion is the responsibility of the readers.

Cut the feet to fit the shoes, or modify shoes to fit the feet

If a person cannot fit their feet into a pair of small shoes, in the end, should they use a small knife to cut their feet so they can fit into the shoes, or is it better to make the shoes big enough to fit the feet?

The answer is obvious.

But some people with narrow original knowledge and thinking can't imagine that the shoes also can be reformed, so they do things literally.

For example, with regard to our Chinese medicine, there are some people who say that traditional Chinese medicine "is not a science."

Here, we only want to use this as an example to explain a problem: any idea, including the definition of science itself, the scientific method itself, should be in constant evolution (i.e. improvement, transformation), and should in no way be adequate to change the shoe.

For some, Chinese medicine theory (meridian assumed), through all available scientific methods, cannot confirm its existence and cannot explain the phenomena associated with it (such as acupuncture meridians and acupuncture anesthesia), so at this time, there are two attitudes: one is that since the existing scientific theory cannot explain it, then you need to discover new scientific theories to enrich the treasury of scientific theory, which will also advance scientific theory itself; if the available scientific method cannot lead to effective means and methods, then it should improve the definition of the scientific method so that the scientific method can also generate awareness of what it ultimately means. This attitude is changing the shoe to fulfill the foot attitude. We believe that the vast majority of medical scientists around the world will adopt this attitude. Even a Millennium Research does not come out; it would have to continue to study even more for a thousand years, never stopping, and this is a push forward in scientific attitude.

Another is the definition of Chinese medicine as "not scientific" by the experts, who think that since the existing scientific theories and methods cannot explain Chinese medicine, then Chinese medicine is not a science. Their attitude is literally a practice; it is to prevent the practice of scientific development and progress because their attitude will not allow the study of Chinese medical scientists to stop their scientific work on TCM!

Still further, since it is so difficult to study the scientific theory of traditional Chinese medicine, once it is indicative of scientific breakthrough, it may break more than just a scientific theory of traditional Chinese medicine, but will lead to a whole level of human health breakthroughs, and even lead to breakthroughs in scientific theories, science and technology, and the scientific method itself.

For example, we discuss in this book the problem of human understanding of the universe. Even traveling for tens of millions of years, people will not get to the edge of one light year from the solar system, but we have chosen to think that we "may be close to the end of our exploration of the ultimate laws of nature." In fact, the Big Bang model of the universe does not make as much sense as the beginning of the world story of thousands of years ago from ancient China.

The Chinese God Pangu appeared and created the universe from chaos. This story, which answers the questions of what existed before and outside of the universe, has all the answers.

On the other hand, the story of the Big Bang begs the question, Big Bang before what? What exploded? Out of what was the Big Bang produced? The discussion has no answer, the typical "to kill without bury!"

On the road of scientific research, changing the shoe to fit the foot is the proper scientific attitude. Only in this manner will it be possible to bring forward our scientific innovation.

Figure 10.2 I also wanted to occupy an Examination Champion – counted it a dream return

Inspiration gotten from the "Eight-portion Stereotyped Essay (八股文)"

Chinese history gives the "Eight-portion Stereotyped Essay" few positive evaluations. The "Eight-portion Stereotyped Essay," as a symbol of the old feudal culture, by violent blow can be described as "notorious."

Baidu Encyclopedia <八股文 Eight-portion Stereotyped Essay> states:

The "Eight-portion Stereotyped Essay" was sinful in history

- The first is that it spoils the reading experience.
- Followed by its lack of practical value.
- Third, it is formally serious.
- Fourth, it proposes nothing new.

The idea that it has been stereotyped in history is not without merit:

- First, scholars were influenced by Confucian ethics to stereotype the study.
- Second, writing theory and techniques for future reference can be stereotyped.
- Again, it provided the text for future generations; the "Eight-portion Stereotyped Essay" was a fine model.
- Last, the "Eight-portion Stereotyped Essay" played a role in fueling some later literary styles, such as couplets of maturity and development.

Seeing comments like that, I cannot help laughing a little. These comments basically can be said to stand on the position of intellectuals from a cultural perspective and ethical point of view to evaluate the Eight-portion Stereotyped Essay.

However, we must not forget that the basic idea of the traditional Chinese is literally "article sell imperial family," and emperors used the Eight-portion Stereotyped Essay as a standardized tool for their selection of management personnel, and it was originally not intended to generate or promote cultural development, so why are we talking about its cultural contributions?

This is just like today's education and examination system, which was originally to fight for possession of educational resources and to set up the selection mechanic of qualified labor; the rest of innovation and education had to step aside.

As a tool for the ruling class to select personnel, the Eight-portion Stereotyped Essay provided so many benefits that we don't even have to think to list a few, but we would not want to waste ink.

What we want to say is that culture and education in China since Confucius and Mencius is primarily aimed at the ruling class training, so there is a strong Chinese cultural official standard atmosphere. The Eight-portion Stereotyped Essay just announced this purpose in a more clearly visible form.

Of course, whether, in fact, the ability to read well can indicate a good management talent is a big question mark.

At least we will not consider a romantic, always impulsive poet who is prone to outbursts capable of management positions in even a small company. For example, it would be difficult to make a good manager out of Li Bai, even though he was the greatest poet. Otherwise, he would not have let people around the emperor for his boots and made the relationship stressful, which we will not discuss here.

Plato wrote *Utopia* to discuss political and educational ideals.

Plato was convinced that for the politics of a country to be an ideal fit, you first have to make education fit the ideal. So he spent half the book's length discussing what kind of education the ruling class needed.

The course he set is very simple: a person should only learn gymnastics and music before the age of twenty, and the rest, if necessary, should be learned in school after the age of twenty. Learn gymnastics, and the body will be strong and fit; learn music, and the soul will be in harmony. Having a fit body and mind in harmony provides most of what you need to do a good job.

But, with the Industrial Revolution and the high efficiency of education and scientific and technological progress, no matter what the origin of the traditional culture of each nation in the world—Plato, Confucius and Mencius, or whatever—by the fatal impact of crisis in the life of the nation, oppressed countries without exception turned to Western education methods, division razed to the barbarians.

Therefore, the main goal of education in its globalized industrial scale mode is to efficiently train skilled workers to provide an adequate, qualified workforce for industrial production, enabling the rapid development of the nation and helping it to win in the competition of country living space.

This goal has not changed; only the technical personnel training target, from the original traditional industries to the information industry, has expanded. When the former U.S. President George W. Bush left the white house, he excused his reasons for launching the war in Iraq as being just a "wrong" information interpretation. While an understatement, the comment well illustrates the background of this goal.

In this globalize mechanical model of unified trained researchers, it is difficult to have hope that their thoughts will jump out of mechanized assembly line mode.

In the prison of the stereotyped format, it is difficult to produce true literature and art; in the steel belt, it is unlikely to plant saplings for a new scientific theory.

If people do not face such a harsh reality, science will eventually become, as John Horgan points out, science ended!

DISCUSSION ABOUT GENIUS — SORROW OF MODERNIZED EDUCATION

If at home you see the bottle flowers made by mother are in full bloom, Xi, Murong has that feeling, "very upset, very anxious, walked beside the flower has been cried: "how do how to do? Face of a small purplish blue flowers with golden eagle tail, our hearts will mess up, do not know how to treat it."; or, as Yu, Guangzhong, as a flower "to watch until desperate before leaving. " If the child has a similar feeling, definitely not use "what nerve" sort of nonsense put his subtle genius nipped in the bud; should be carefully nurtured, in fact, very well pretend inadvertently encourage, train, so he grows up naturally continually. Casual neglect, ridicule childish sharp, are fatal; and too much praise will be like too thick ammonia fertilizer, injury seedling genius.

If the child has feelings similar to a keen poet, but has to learn physics and chemistry as a scientist, he might be able to become a mediocre, competent researcher, or perhaps he will be totally disappointing. Mathematics did not have sense; literary genius was buried and smothered.

Maxima often, the Bole does not often. Finding genius is often a very difficult thing.

Usually our common method for identifying talent is to see how much a child can absorb knowledge. If a child can finish the full course of primary school in one or two years, less than half the time people take to finish high school, we basically conclude that he is a genius. Methods such as admission of USTC for Youth Genius Class after the Cultural Revolution is: a unified written, try college class, IQ measurement, comprehensive investigation, or to the amount of already acquired knowledge, and ability to absorb knowledge, as the basic selection standards. And in the training of selected geniuses basically there is also nothing new.

In fact, this is not necessarily a completely correct method for identifying talent. More questionable is the need to cultivate genius. Such trained talent, in general let talented young people can serve the community in advance of several years only. Even at a young age to become president of a well-known company, or even become relatively young famous professor, it is nothing but obliterate all one only.

We have kept a 2005 Guizhou Metropolis Daily newspaper from December 6, which reported in section A2 the interesting news that "Female prodigy home educated triggered war." The next day, the news reported (section A3) that "Happy growth is more important than the 'child prodigy.'" From the whole story, we can find some interesting things to discuss. A nine-year-old girl had attended second grade in middle school. Her father insisted on continuing her education at home, and her mother and grandparents wanted her return to school. The father said: "Our daughter is not a child prodigy, and our approach to education is not simple." Their teacher said that she was not a particularly clever child who barely kept up in the first grade, but had found it hard to catch up in the second grade in the middle school.

Absorbing knowledge at a speed beyond the ordinary, of course, can be regarded as a sign of superior intelligence. Such an ability can help people to complete their studies or work sooner than other people, or to participate in study at an earlier age, but not necessarily to ultimately do better than ordinary people. Such a genius shortens their time spent in school, ahead of contributing to society, and can advance to become useful to society with outstanding talent but not necessarily be able to make a seminal contribution to mankind. Like the above mentioned female child prodigy, in fact, many children are able to achieve this kind of learning speed and to absorb the level of knowledge, mainly based on the conditions in which the children learn. In addition, some intelligent people mature early, and some people late. Those who mature early may learn faster than others but may not necessarily have more advantages over others after growing up. The USTC Youth Class deliberately nurtures young talent, encouraging them to move into society as soon as possible to contribute earlier, and there is no essential difference from the ordinary undergraduates.

Another is the genius who has the ability to create. Such people are more difficult to find, almost no chance to enter the USTC Youth Class such special training base. Such people are often only interested in a particular field or something in which they can exhibit a particularly extraordinary intelligence. They can often bring a revolutionary breakthrough in the field. But they are more likely to be neglected socially or suffer an abrasive torrent turning them into the same rounded pebbles as everyone else.

And what we absolutely need is that creative genius who will have the ability to shatter the front edge.

For example, a five-year-old, Dou Kou, began to publish works and should be considered an extraordinary genius. Online, in early 2009, his introduction went as follows:

Dou Kou, at four and a half years of age began to write a diary, and he has already written 24 notebooks covering 3000 days. He expanded his writing from the ordinary diary to inspiration diary and then to fairy tales, science fiction and literature stories, and lengthy monographs. When he was five years old, he began to publish poems and fairy tales and has published more than 30 articles. As a six-year-old, he wrote and published a lengthy autobiography *Dou Kou Wandering Mind*, at the age of eight, he published *Dou Kou Years*, and, at ten, he published the novel *Childhood's Eyes*. Today, 12-year-old Dou Kou has created and completed a full-length literary work *Alas Juvenile Days* (to be published).

Dou Kou went to school and directly attended the fifth grade at six and a half years of age, and was intermittently in the school for three years. Due to Dou Kou's language skills and strong understanding of other subjects, in June 2006, he was less than 12 years old when he finished all primary and secondary school studies and was a high-school graduate in Shanghai. While at school, this little young man has a deep sentiment in the current education system, after many active thinking, wrote the exploratory research book *Declaration on Education Revolution*.

It is said that, because of tuition, he went to Indonesia to study. We were happy for him but also sorry for the public. Is he not a rare creative genius?

(Several old examples show how many years it took for this book to be written.)

Sunday, I went to the Berkeley Kunqu opera club, and I remember that Mrs. Conghe Zhang had just passed away. At the age of 17, she had a test in the Beijing University. In Chinese, she got full score, but in math, zero. The Chinese Literature Department at Peking University cherishes talented people, and they broke school rules and gave her exceptional admission. Later, she also made brilliant achievements in calligraphy, poetry, opera, and other fields. She was calligraphy professor at Yale University, teaching Kunqu opera for many years, and had made outstanding contributions for the dissemination of Chinese culture.

When Wu Xiao was turning five years old, I taught him math. I remember very clearly that it was the night after we moved to a new apartment near the school.

"5 - 3 = 2," I said.

"3 - 5 is equal to how much?" Wu Xiao immediately asked without thinking.

I was stunned, and I looked at him for a long time, speechless. Ha-ha, we get the gems, our son is a genius.

Wu Xiao quickly finished elementary school arithmetic. So we taught him to use a Chinese junior-high-school algebra textbook. Sometimes we spoke to him about the concept of infinity and proof points infinitesimal and the like, and he could comprehend immediately.

However, trouble came. After learning so much math, math class in school, of course, was no fun for Wu Xiao. So he did not want to attend lecture, lack of discipline and other issues were out. When we were busy with work and study, there was no good time to care about the children, and I did not know in the end how to handle such a situation. The children mathematics were far ahead, while in the other disciplines, especially English, it was difficult to achieve the same advanced level.

Later, when he was in fourth grade, we decided that Wu Xiao would no longer study mathematics in his spare time (he had finished junior-high-school algebra using the Chinese textbook), and let him study music, sports, and so on that had nothing to do with school stuff. Fortunately, Wu Xiao had mastered the method of self-learning any subject, which he'd learned during family teaching over the previous few years. In high school, his mathematics improved through his own efforts. He went to the university to listen to math class. Because we did not use coercion to stifle his interest in learning!

I often imagine if we had been able to do something, what would have happened if we'd taught Wu Xiao how to create a learning environment for himself. Although Wu Xiao in real life developed very well, I'm sure we have personally wasted a genius in some sense, though this was due to the circumstances of life. If we were able to present my current vision on Wu Xiao's education, he would be likely to become a man planting trees and not just a tree-climbing expert vertically jumping and flying up his predecessors' planted trees.

With the efforts of an individual or family, it is impossible to modernize education from the production line. And once a person has been included in the modernization of the education production line, he loses the courage and even desire to jump from education automatic production line, to put forth any effort to find a green land and plant a new tree.

This is a sad human modernization of education!

On a recommendation to China's college entrance examination – Start from the reform of test papers

This year saw more false students, cheating and so on, negative news reports about the college entrance examination; it was not too happy. A simple technical means you can basically put an end to the phenomenon, but letting it occur annually makes it a social ill.

For this reason, I sincerely give recommendations on changes to the college entrance examination test paper design to heads in charge of education. A new test paper, to a certain extent, can change the idea that getting into college is the only goal of education, can greatly achieve the purpose of selecting for different talents, and can basically eliminate the phenomenon of cheating in exams.

This is called fighting fire with fire, using the test to bring down the rigid examination, and allow the examination to achieve the purpose of the guide!

First, the number of questions has to break the routine. If the test time is two hours, then the new amount should be at least three hours of questions and even four hours for the candidates to complete. This is because, in such a test, the first test can be agility of thinking. Of course, the massive number of testing problems is not this only purpose.

Q & A on the exam can be divided according to the needs of people, and the exam focus should be different for the various subject groups: there will be a basic skills test, but also one for technique, a test on ways of thinking, and also IQ test, and so on. The classification of the candidates is transparent. The result is that knowledge is widely covered, and candidates can choose the type of questions to learn according to their own preferences. When the candidates learn their favorite things, they are actively rather than passively learning.

Thus, the results of the candidate's subject matter and their answers can be used to choose the right people.

For example, with computer science, students can choose those logical thinking questions. MBA students can elect to have their comprehensive knowledge probed not in very depth. For students with research talent, they can elect to answer those difficulty problems, and so on.

Some problems can test the level of flexibility of handling knowledge, some problems, can test the candidate's creativity, and so on.

Although during the initial time there will be some of the questions, need the run-time problem-solving process, but will be much better in that chosen specialty talents fit with their future professional development. Candidates can work in their favorite areas of expertise, and it will be more conducive to the social demand for an excellent workforce.

So, a lot of knowledge covering a wide range of test papers, but this means that the school educational system needs to change. Teachers are not clinging to a few questions and going back and forth between several sets of textbooks and testing skills. Students can follow their own interests, swim in the ocean of knowledge. The learning atmosphere that follows will inevitably play a huge change.

This is the use of exam-oriented education to correct the shortcomings of the original, and turn it into a driving force for the candidates' proactive and flexible learning.

In order to reduce cheating, the topic of computer software can do the arbitrary grouping of papers, where each subject has small changes, and the order of the answers are not the same. Each examination paper can be numbered so that the packet is a random number. Thus, papers with the same difficulty have substantially different answers, in addition to counterfeit than replace other ways of cheating candidates, other basic avoided. And require candidates to press their fingerprint on the paper's corner so you can use the personal computer to check the authenticity of the candidates. This year Wuyuan scenic spots tickets have been required to press the fingerprints.

Further, if we also want to eliminate the test paper leak problem during transportation, a GPS can be installed in boxes for the papers so that they can be tracked from packing to unpacking, under full supervision.

These simple questions with little change and technological means, we can achieve many purposes, why not do it?

China a unique opportunity

Although there are some unpleasant dark sides to China's education, it can still be said to meet a unique historical opportunity.

To create a unique and appropriate planting of saplings to grow in the current environment of social conditions, China has its unique advantages.

Because it practices socialist democratic centralism, the state has greater freedom and disposable resources, and people are willing to put their children into more advanced pilot projects. Therefore, we can design a relatively free environment, seek some talented children from an early age, and educate them in a unified culture that is different from the mechanized mode of educational methods. We hope that these children grow up in a relatively natural environment and can ultimately create new miracles for humanity. Considering all aspects, I am afraid that these are difficult things to do in other countries.

Of course, this is just our little childish imagination. Real-life experiments are probably not a very easy thing. Not only do they need a strong state support, but they also have a long experimental period.

However, even this is nothing to imagine. Considering the various conditions specific to implementation, then our country can make the probability of success much greater than in other countries.

This is a unique performance advantage of Chinese socialism. Maybe we can seize this opportunity?

Unfinished words

Breath wrote here, it is not enough, simply just begun.

There is so much to say to the students, the parents, and to the teachers. We want to write a book for each of you. Unfortunately, counting with fingers, the project is too vast.

Our society is entering an exciting period of major transition. This period coincides with the development of thought and social development so that students can be in harmony with the educational system. This is the sacred mission entrusted by the times, giving rise to the historical opportunity of the Chinese nation for centuries of the best opportunities. Each of us plays an indispensable role.

For the future of the Chinese nation, please do that duty!

CHAPTER 11

MATHEMATICS IS OLD

Mathematics is old

The first chapter title of my book *Who Should Talk* About *Cosmos* is "Is Math Old?" At that time, I intuitively felt that mathematics was old. Mathematical creativity was gradually fading.

Mathematics is old.

It is now my understanding, after 10 years thinking.

Today, after we exchanged our views with more friends, after we and some mathematicians exchanged views, I think that "mathematics is old" is a basic fact.

A fact that is at stake.

Mathematics is the foundation of natural science, a symbol of natural science.

If you say that mathematics is old, does that mean that natural science is old? Are humans no longer able to create revolutionary breakthrough inventions and discoveries? Is it like Hogan said in his book *The End of Science*: "Today, science has revealed the basic facts of man and the universe, from the nature of scientific work has been completed on the terms in almost all 'pure science', the major discoveries have been found, and no major incidents of amazing."

There is a series of ideas worth exploring.

First, is mathematics really old?

If mathematics is really old, why is it old?

If mathematics is really old, is science old?

Do we have a way to make mathematics rejuvenated?

CLIMBING TREES AND PLANTING TREES — IS MATHEMATICS OLD?

We often do not know what the real reason is that makes people old.

As a person ages, his mental life is gradually reduced, his pace and external performance increasingly falters, and his actions become more and more sluggish.

In fact, why aging is a difficult problem is tough to say clearly.

The same molecules make up our cells; why are these are "old" while those are still young? On the molecular level, what kind of mechanism it is?

The idea that "math is old" also is also inferred from its external performance.

A few manifestations of the aging of mathematics:

After simple mathematical problems are gradually resolved, mathematics is powerless in the face of the complex world. The delicacy of mathematics leaves it powerless in front of the changing, rough, real world. Reluctant to use sophisticated mathematical tools to describe the wild and varied system, only is a fine play on practical system! Only free universe left a space to play with exquisite mathematical formula, the rest are bluntly replaced by computer numerical analysis.

Let's look at a little story.

The protagonists of the story are the coach and the student.

The story takes place in a forest.

The trees of this forest were planted some long number of years ago, some hundreds of years ago, and the youngest is one hundred years old. Some trees were planted by the elderly, but some were also planted by people less than twenty years old at that time.

The coaches begin to teach the students at a very young age.

This is a modern, extremely formal education.

When they are young, the coaches tell the students how great, how excellent are those who planted a tree in this forest, how talented they were. Slowly, the coaches teach the students to look on these trees in awe.

As they grew older, the coaches started teaching them tree-climbing skills.

These trees have a rich fruit of inexhaustible wisdom.

Some children are not good at climbing trees; they do not have tree-climbing qualities. They are still encouraged, but after multiple failures, they are sent for other training. These students spend the rest of their lives overlooking the forest from afar, with respectful or disgusted moods.

Some children are very clever, with very good tree-climbing abilities, and they are left to slowly climb this or that tree.

There are exotic flowers and fruit trees, with long branches, vines attached to trees and entangled with them. The children grow up in the world of this forest, climbing higher and farther in the branches. They build houses in the trees, build sheds, and connect trees with vines, tireless, hard work, until they have unconsciously grown old!

Year after year, the size of the forest is the same, and for the last hundred years, no more trees have been added.

Generation after generation, no one thinks to go down from the trees to plant a new tree.

The forest is mathematics; the trees of the forest are the branches of mathematics founded by the predecessors: this one is thousands of years of geometry, that there is a long history of algebra. Here is a little, younger tree for group theory, and there is the small, young Boolean algebra tree.

Almost every tree was planted at least a century ago by math masters.

In the past century, mankind was busy climbing trees only!

Humanity over the past century has not planted a new tree!

Humanity over the past century has not created a new mathematical discipline! (We asked some mathematicians on this issue.)

Figure 10.3 is the cover of "A History of Mathematics."

In this book, which is more than 700 pages thick—published in 1968, reprinted in 1989 and 1991—after 1900, the recorded history of mathematics and major groundbreaking mathematical achievement is blank.

Is humanity getting more and more stupid?

Have humans failed to produce a talent in the last hundred years?

Of course not!

Why is mathematics old? Is it due to natural factors or human factors?

If mathematics is old, is science also old?

Do we have a way to make mathematics rejuvenated?

Can we let science never age?

We can study a number of areas, including educational, economic, moral, cultural, and even science itself, from the status quo, from history, from various development angles, etc., to seek answers.

The aging of mathematics is due to human creativity aging; and the aging of human creativity and human teaching methods, or educational purposes, are inextricably linked. The next part of the book focus on the problem. We will combine education and creativity training with some superficial discussion of establishment, etc. for the purpose of shallow discussion.

During the Cultural Revolution, I wrote a little piece of comedy, "The Third Shoot" for a propaganda team in the countryside. Because of its humor, it won a great success. But then I went on looking for more humor material for a new comedy, no matter how hard I looked, I could not find a new spot to laugh. In retrospect, I realized that people did not smile in their thought of those years. It is hard to break thought to write interesting pieces.

Figure 10.3 Cover of the book "A History of Mathematics". In this more than 700-page book, published in 1968, and reprinted in 1989 and 1991, is the history of mathematics. After 1900, the recorded history of mathematics and mathematical achievement is blank

Then, in the unity of thinking people who grew up in the environment, when they do not even sense the invisible ideological imprisonment, how can they think of breaking the bounds? Only had lamented "Subject Mathematics invention have been finished."

Forest Guardians

In the mathematics forest, the tree of "algebra" is already a millennium old, and that young tree "Boolean algebra" has just come out for a while. In this forest, most are unaware that there are guardians. They are familiar with every tree, and every tree they love carefully. They enthusiastically introduce people to each tree, and they are keener to train the children to climb the trees.

When did those guardians actually appear?

It is not possible to pinpoint a specific time, but the approximate time is when the Industrial Revolution was basically completed. They brought the mechanization of the Industrial Revolution into this forest, and began to manage it with mechanized efficiency, making full use of the forest for human wealth increase. Thus, a myriad of applications was produced from the forest produce, and dazzling technological achievements emanated from there.

However, the guardians only concentrate on caring for each tree, seeking only to guide each child to climb a tree. They only allow children to climb trees near the forest; they only encourage children to climb the tree to pick fruit. They unintentionally push similar, industrialized standards to educate their children, and the children train in an almost uniform way of thinking. So not a single child has the idea of planting a tree, though even if someone were to occasionally throw out the idea, the guardians, who set all the rules, would filter the idea out.

For the last one hundred years, no new tree has appeared in the forest.

How can humans face such a dilemma?

How can human beings be able to, one after another, plant new trees?

Here, I just thrown out some rubble, made from my ability to see, my ability to think of problems and phenomena, and to try to give limited answers. I only hope that this piece of debris can arouse a few ripples of shock and have some resonance in the thoughts of interested people so that they can fuse together to seek possible answers and solutions.

This is disturbing the status quo, but also an exciting challenge. The rise of science and technology is our opportunity: perhaps new saplings of mathematics and scientific theory will grow into forests cultivated by the Chinese younger generation!

This is possible as long as we can grasp the opportunities of education to make bold innovation.

"END OF SCIENCE" AND THE RISING OPPORTUNITIES OF SCIENCE AND TECHNOLOGY

In 1997, *Scientific American* senior writer John Horgan published a book entitled *The End of Science*, after visiting scientific leaders in more than 10 fields. The book presents the following idea: today, science has revealed to man an understanding of the basic facts of the universe, and scientific work discovering the big concepts has ended. In almost all the "pure sciences," the major discoveries have been found, and there will be no more amazing revelations.

In 2006, on Mid-Autumn Festival day, I queued in payment line of Berkeley Bowel grocery store and picked up a *Discover* magazine reading, when, unexpectedly, saw John Horgan, after one decade, picked up the old topic in an article entitled "The Final Frontier." In the article, he said: "Over the past decade, scientists have announced countless discoveries that seem to undercut my thesis: cloned mammals (starting with Dolly the sheep), a detailed map of the human genome, a computer that can beat the world champion in chess, brain chips that let paralyzed people control computers purely by thought, glimpses of planets around other stars, and detailed measurements of the afterglow of the Big Bang. Yet within these successes there are nagging hints that most of what lies ahead involves filling in the blanks of today's big scientific concepts, not uncovering totally new ones."

I was surprised that, a decade later, there was not even a single scientist who could convince Horgan that his science termination assertion was wrong, but Horgan after a decade of thinking was still convinced of his point of view, that, for a plurality of fields, there was overwhelming proof of the end of science.

His main argument is that, in all areas of pure scientific theory—mathematics, physics, chemistry, etc.—there has not been any innovation. In other words, in all scientific disciplines, in the scientific forests, he did not see any people today planting a tree!

Like Horgan, we also believe that the progress of scientific theories is actually at a standstill, despite the dizzying variety of new applications of science and technology and discovery; in fact, they are just filling in and completing the existing scientific theoretical gaps, but there is no fundamental basic scientific theory that has made a breakthrough.

Mathematics has not produced any new branch, although the number of mathematics researchers today has increased hundreds of thousands of times; there have also been no new revolutionary scientific theories, although number of scientific researchers has increased one hundred million times.

Figure 10.4 Horgan published in *Discover* magazine the article "The Final Frontier"

Recent human scientific achievements have been focused on getting fruit from the trees planted in previous efforts, and not on planting new trees.

However, we do not believe that the pause in theory creation is because of humans have reached the end of science. As long as we dare to face the fact and to find the reason that humans lost the ability to plant trees, we might have new space in the world for the cultivation of a new forest!

SOLEMN CLASSICAL SCIENCE AND FRIVOLOUS MODERN SCIENCE

I do not know when it began, but science is becoming increasingly non-serious, floating noise.

There are many man-made factors, such as the tendency in school to talk about right and wrong but not about truth. And this ugly phenomenon is derived from the complexity of modern science itself, and the difficulty of using black and white to determine yes or no with non-ambiguity, and more insight into human thought, which is all mixed arrogance and does not recognize human does have its exhaustive.

The first is to treat the attitude of the speed of light.

Einstein asserted that the speed of light is constant. Michelson–Morley and many similar experiments "proved" this assertion. However, if the light is so fine that their quality is only one millionth of elementary particles or ten millionth, what means humans can distinguish between details of the light? If the light is transmitted in units of ten thousands of light-years of distance then start getting "tired", humans can have any means to simulation light travel in unit of ten thousand light-years? In the Michelson–Morley experiment and other similar experiments, the experimental time duration of light traveled maybe even less than a year or less than a light-year distance, right? It is not possible to simulate a hundred light-years or more in lab.

Previously, we discussed in detail that the stars observable by humans is more than one photon statistics, can be observed in a relatively close to the radius will be "tired" and this tiredness can be observed.

Another example is "dark matter" theory. Is this thing in itself a very strange thing with a very unnatural theory? I have a bet with NASA because I basically do not think that dark matter exists. We'll see what kind of magical means cosmologists use to find or create a little dark matter. If it has to account for 96% of the matter in the universe, it can never be dark matter. The reason for this, I have fully discussed earlier in the book.

The application progress of modern technology covered up the lack of theoretical innovation, and the power of capital driven talents tends money accumulation place. So willing to engage in some research fields with little research funds are essentially rare. In this case, an amateur's illusory research of the universe is simply surprising!

The first draft of this book, my friend had forwarded to an editor. His comment was: if this man can write such boring stuff for a decade, he is definitely an eccentric.

I listened to the remark and could only respond with a wry smile. I feel very strange myself, ah! I spent a decade of my spare time on this and did not earn a penny—sick, ah? What kind of world view is that? What kind of universal view? Can you be more stuck in reality?

My glorious image is ruined by this book, ah!

In *The End of Science*, the first example HORGAN used was cosmological science! Why was cosmological science the primary basis of his argument? Should we not consider carefully? Who impeded the progress of space science and development?

Einstein's story about go back to the time he was born to kill his father turned science into scientific fiction fantasy, so in "Who Should Talk about Cosmos" the relativity theory was named " Relativity Literature."

Based on elementary-school arithmetic, the relativity concept was established. Is it not then a few famous scientists who say, "There are only a few people who can understand," and who cut it out of the emperor's new clothes? If you cannot see the emperor's new clothes, it means your eyes have problems. Hurry to cure your eyes.

I think that discerning right and wrong in the view points of universe is more meaningful to the human progress. If our work can give the field of rigid untouchable produce little impact, so that people realize that since a hundred years the human predicament that no new trees can be cultivated. So by the discussion of the end of science topic that can impact authority, people can realize the plight, to have excited desires that want to jump down from a trees to plant a new tree and build new forest. It will be a wonderful thing!

We have repeatedly stressed: our hopes in the younger generation, hope in the next generation. We do not let them grew up to become only slavishly follow the master's authority and conformist students. We want to guide them to create new theories and cosmology great standard-bearer.

We do not need science and technology God that imprison scientific progress!

We call for God who does creation that advancing the science!

Chapter Summary

Aging is a slow, subtle process, particularly light during adolescence. Years can only unconsciously, the body's cells decay, erosion of personal life. Only until is shocked to find that time has painted one's temples with frost, then was woken up life is so short.

Compared to humans, any one particle of the universe belongs to eternity!

The mechanisms of modern society have used up the land available for planting.

Thanks to the rise of the network, so that human knowledge can be shared beyond time and space, in this era, there is likely to grow master genius so that the century of mathematical decline may be stopped now.

Unfortunately, the network has become a major social tool, and only a handful of strong-willed young people, not captive by neither the social networks nor the gaming network. And these people are not necessarily really geniuses.

So the mediocre becomes the mainstream of the society, and for universal social communication, pleasure becomes fashion. In humans, this pleasure of mediocrity leads us towards a more mediocre existence.

Ugh!

Mathematics is old!

APPENDIX: ON THE ELECTRODYNAMICS OF MOVING BODIES

By A. Einstein June 30, 1905

(Where the underline delineates the part of the criticism we quoted in the article. Please check whether it is changed Einstein's original meaning.)

It is known that Maxwell's electrodynamics—as usually understood at the present time—when applied to moving bodies, leads to asymmetries which do not appear to be inherent in the phenomena. Take, for example, the reciprocal electrodynamic action of a magnet and a conductor. The observable phenomenon here depends only on the relative motion of the conductor and the magnet, whereas the customary view draws a sharp distinction between the two cases in which either the one or the other of these bodies is in motion. For if the magnet is in motion and the conductor at rest, there arises in the neighborhood of the magnet an electric field with a certain definite energy, producing a current at the places where parts of the conductor are situated. But if the magnet is stationary and the conductor in motion, no electric field arises in the neighborhood of the magnet. In the conductor, however, we find an electromotive force, to which in itself there is no corresponding energy, but which gives rise—assuming equality of relative motion in the two cases discussed—to electric currents of the same path and intensity as those produced by the electric forces in the former case.

Examples of this sort, together with the unsuccessful attempts to discover any motion of the earth relatively to the "light medium," suggest that the phenomena of electrodynamics as well as of mechanics possess no properties corresponding to the idea of absolute rest. They suggest rather that, as has already been shown to the first order of small quantities, the same laws of electrodynamics and optics will be valid for all frames of reference for which the equations of mechanics hold good.[1] We will raise this conjecture (the purport of which will hereafter be called the "Principle of Relativity") to the status of a postulate, and also introduce another postulate, which is only apparently irreconcilable with the former, namely, that light is always propagated in empty space with a definite velocity c which is independent of the state of motion of the emitting body. These two postulates suffice for the attainment of a simple and consistent theory of the electrodynamics of moving bodies based on Maxwell's theory for stationary bodies. The introduction of a "luminiferous ether" will prove to be superfluous inasmuch as the view here to be developed will not require an "absolutely stationary space" provided with special properties, nor assign a velocity-vector to a point of the empty space in which electromagnetic processes take place.

The theory to be developed is based—like all electrodynamics—on the kinematics of the rigid body, since the assertions of any such theory have to do with the relationships between rigid bodies (systems of co-ordinates), clocks, and electromagnetic processes. Insufficient consideration of this circumstance lies at the root of the difficulties which the electrodynamics of moving bodies at present encounters.

I. KINEMATICAL PART

§ 1. Definition of Simultaneity

Let us take a system of co-ordinates in which the equations of Newtonian mechanics hold good.[2] In order to render our presentation more precise and to distinguish this system of co-ordinates verbally from others which will be introduced hereafter, we call it the "stationary system."

If a material point is at rest relatively to this system of co-ordinates, its position can be defined relatively thereto by the employment of rigid standards of measurement and the methods of Euclidean geometry, and can be expressed in Cartesian co-ordinates.

If we wish to describe the *motion* of a material point, we give the values of its co-ordinates as functions of the time. Now we must bear carefully in mind that a mathematical description of this kind has no physical meaning unless we are quite clear as to what we understand by "time." We have to take into account that all our judgments in which time plays a part are always judgments of *simultaneous events*. If, for instance, I say, "That train arrives here at 7 o'clock," I mean something like this: "The pointing of the small hand of my watch to 7 and the arrival of the train are simultaneous events."[3]

It might appear possible to overcome all the difficulties attending the definition of "time" by substituting "the position of the small hand of my watch" for "time." And in fact such a definition is satisfactory when we are concerned with defining a time exclusively for the place where the watch is located; but it is no longer satisfactory when we have to connect in time series of events occurring at different places, or—what comes to the same thing—to evaluate the times of events occurring at places remote from the watch.

We might, of course, content ourselves with time values determined by an observer stationed together with the watch at the origin of the co-ordinates, and co-ordinating the corresponding positions of the hands with light signals, given out by every event to be timed, and reaching him through empty space. But this co-ordination has the disadvantage that it is not independent of the standpoint of the observer with the watch or clock, as we know from experience. We arrive at a much more practical determination along the following line of thought.

If at the point A of space there is a clock, an observer at A can determine the time values of events in the immediate proximity of A by finding the positions of the hands which are simultaneous with these events. If there is at the point B of space another clock in all respects resembling the one at A, it is possible for an observer at B to determine the time values of events in the immediate neighbourhood of B. But it is not possible without further assumption to compare, in respect of time, an event at A with an event at B. We have so far defined only an "A time" and a "B time." We have not defined a common "time" for A and B, for the latter cannot be defined at all unless we establish *by definition* that the "time" required by light to travel from A to B equals the "time" it requires to travel from B to A. Let a ray of light start at the "A time" t_A from A towards B, let it at the "B time" t_B be reflected at B in the direction of A, and arrive again at A at the "A time" t'_A.

In accordance with definition the two clocks synchronize if

$$t_B - t_A = t'_A - t_B.$$

We assume that this definition of synchronism is free from contradictions, and possible for any number of points; and that the following relations are universally valid:—

1. If the clock at B synchronizes with the clock at A, the clock at A synchronizes with the clock at B.

2. If the clock at A synchronizes with the clock at B and also with the clock at C, the clocks at B and C also synchronize with each other.

Thus with the help of certain imaginary physical experiments we have settled what is to be understood by synchronous stationary clocks located at different places, and have evidently obtained a definition of "simultaneous," or "synchronous," and of "time." The "time" of an event is that which is given simultaneously with the event by a stationary clock located at the place of the event, this clock being synchronous, and indeed synchronous for all time determinations, with a specified stationary clock.

In agreement with experience we further assume the quantity

$$\frac{2AB}{t'_A - t_A} = c,$$

to be a universal constant—the velocity of light in empty space.

It is essential to have time defined by means of stationary clocks in the stationary system, and the time now defined being appropriate to the stationary system we call it "the time of the stationary system."

§ 2. On the Relativity of Lengths and Times

The following reflexions are based on the principle of relativity and on the principle of the constancy of the velocity of light. These two principles we define as follows:—

1. The laws by which the states of physical systems undergo change are not affected, whether these changes of state be referred to the one or the other of two systems of co-ordinates in uniform translatory motion.

2. Any ray of light moves in the "stationary" system of co-ordinates with the determined velocity c, whether the ray be emitted by a stationary or by a moving body. Hence

$$\text{velocity} = \frac{\text{light path}}{\text{time interval}}$$

where time interval is to be taken in the sense of the definition in § 1.

Let there be given a stationary rigid rod; and let its length be l as measured by a measuring-rod which is also stationary. We now imagine the axis of the rod lying along the axis of x of the stationary system of co-ordinates, and that a uniform motion of parallel translation with velocity v along the axis of x in the direction of increasing x is then imparted to the rod. We now inquire as to the length of the moving rod, and imagine its length to be ascertained by the following two operations:—

(a)

The observer moves together with the given measuring-rod and the rod to be measured, and measures the length of the rod directly by superposing the measuring-rod, in just the same way as if all three were at rest.

(b)

By means of stationary clocks set up in the stationary system and synchronizing in accordance with § 1, the observer ascertains at what points of the stationary system the two ends of the rod to be measured are located at a definite time. The distance between these two points, measured by the measuring-rod already employed, which in this case is at rest, is also a length which may be designated "the length of the rod."

In accordance with the principle of relativity the length to be discovered by the operation (a)—we will call it "the length of the rod in the moving system"—must be equal to the length l of the stationary rod.

The length to be discovered by the operation (b) we will call "the length of the (moving) rod in the stationary system." This we shall determine on the basis of our two principles, and we shall find that it differs from l.

Current kinematics tacitly assumes that the lengths determined by these two operations are precisely equal, or in other words, that a moving rigid body at the epoch t may in geometrical respects be perfectly represented by *the same* body *at rest* in a definite position.

We imagine further that at the two ends A and B of the rod, clocks are placed which synchronize with the clocks of the stationary system, that is to say that their indications correspond at any instant to the "time of the stationary system" at the places where they happen to be. These clocks are therefore "synchronous in the stationary system."

We imagine further that with each clock there is a moving observer, and that these observers apply to both clocks the criterion established in § 1 for the synchronization of two clocks. Let a ray of light depart from A at the time[4] t_A, let it be reflected at B at the time t_B, and reach A again at the time t'_A. Taking into consideration the principle of the constancy of the velocity of light we find that

$$t_B - t_A = \frac{r_{AB}}{c - v} \quad \text{and} \quad t'_A - t_B = \frac{r_{AB}}{c + v}$$

where r_{AB} denotes the length of the moving rod—measured in the stationary system. Observers moving with the moving rod would thus find that the two clocks were not synchronous, while observers in the stationary system would declare the clocks to be synchronous.

So we see that we cannot attach any *absolute* signification to the concept of simultaneity, but that two events which, viewed from a system of co-ordinates, are simultaneous, can no longer be looked upon as simultaneous events when envisaged from a system which is in motion relatively to that system.

§ 3. Theory of the Transformation of Co-ordinates and Times from a Stationary System to another System in Uniform Motion of Translation Relatively to the Former

Let us in "stationary" space take two systems of co-ordinates, i.e. two systems, each of three rigid material lines, perpendicular to one another, and issuing from a point. Let the axes of X of the two systems coincide, and their axes of Y and Z respectively be parallel. Let each system be provided with a rigid measuring-rod and a number of clocks, and let the two measuring-rods, and likewise all the clocks of the two systems, be in all respects alike.

Now to the origin of one of the two systems (k) let a constant velocity v be imparted in the direction of the increasing x of the other stationary system (K), and let this velocity be communicated to the axes of the co-ordinates, the relevant measuring-rod, and the clocks. To any time of the stationary system K there then will correspond a definite position of the axes of the moving system, and from reasons of symmetry we are entitled to assume that the motion of k may be such that the axes of the moving system are at the time t (this "t" always denotes a time of the stationary system) parallel to the axes of the stationary system.

We now imagine space to be measured from the stationary system K by means of the stationary measuring-rod, and also from the moving system k by means of the measuring-rod moving with it; and that we thus obtain the co-ordinates x, y, z, and ξ, η, ζ, respectively. Further, let the time t of the stationary system be determined for all points thereof at which there are clocks by means of light signals in the manner indicated in § 1; similarly let the time τ of the moving system be determined for all points of the moving system at which there are clocks at rest relatively to that system by applying the method, given in § 1, of light signals between the points at which the latter clocks are located.

To any system of values x, y, z, t, which completely defines the place and time of an event in the stationary system, there belongs a system of values ξ, η, ζ, τ, determining that event relatively to the system k, and our task is now to find the system of equations connecting these quantities.

In the first place it is clear that the equations must be *linear* on account of the properties of homogeneity which we attribute to space and time.

If we place $x'=x-vt$, it is clear that a point at rest in the system k must have a system of values x', y, z, independent of time. We first define τ as a function of x', y, z, and t. To do this we have to express in equations that τ is nothing else than the summary of the data of clocks at rest in system k, which have been synchronized according to the rule given in § 1.

From the origin of system k let a ray be emitted at the time τ_0 along the X-axis to x', and at the time τ_1 be reflected thence to the origin of the co-ordinates, arriving there at the time τ_2; we then must have $\frac{1}{2}(\tau_0 + \tau_2) = \tau_1$, or, by inserting the arguments of the function τ and applying the principle of the constancy of the velocity of light in the stationary system:—

$$\frac{1}{2}\left[\tau(0,0,0,t) + \tau\left(0,0,0,t + \frac{x'}{c-v} + \frac{x'}{c+v}\right)\right] = \tau\left(x',0,0,t + \frac{x'}{c-v}\right).$$

Hence, if x' be chosen infinitesimally small,

$$\frac{1}{2}\left(\frac{1}{c-v} + \frac{1}{c+v}\right)\frac{\partial \tau}{\partial t} = \frac{\partial \tau}{\partial x'} + \frac{1}{c-v}\frac{\partial \tau}{\partial t},$$

or
$$\frac{\partial \tau}{\partial x'} + \frac{v}{c^2 - v^2}\frac{\partial \tau}{\partial t} = 0.$$

It is to be noted that instead of the origin of the co-ordinates we might have chosen any other point for the point of origin of the ray, and the equation just obtained is therefore valid for all values of x', y, z.

An analogous consideration—applied to the axes of Y and Z—it being borne in mind that light is always propagated along these axes, when viewed from the stationary system, with the velocity $\sqrt{c^2 - v^2}$ gives us

$$\frac{\partial \tau}{\partial y} = 0, \frac{\partial \tau}{\partial z} = 0.$$

Since τ is a *linear* function, it follows from these equations that

$$\tau = a\left(t - \frac{v}{c^2 - v^2}x'\right)$$

where a is a function $\phi(v)$ at present unknown, and where for brevity it is assumed that at the origin of k, $\tau = 0$, when $t = 0$.

With the help of this result we easily determine the quantities ξ, η, ζ by expressing in equations that light (as required by the principle of the constancy of the velocity of light, in combination with the principle of relativity) is also propagated with velocity c when measured in the moving system. For a ray of light emitted at the time $\tau = 0$ in the direction of the increasing ξ

$$\xi = c\tau \text{ or } \xi = ac\left(t - \frac{v}{c^2 - v^2}x'\right).$$

But the ray moves relatively to the initial point of k, when measured in the stationary system, with the velocity $c - v$, so that

$$\frac{x'}{c - v} = t.$$

If we insert this value of t in the equation for ξ, we obtain

$$\xi = a\frac{c^2}{c^2 - v^2}x'.$$

In an analogous manner we find, by considering rays moving along the two other axes, that

$$\eta = c\tau = ac\left(t - \frac{v}{c^2 - v^2}x'\right)$$

when

$$\frac{y}{\sqrt{c^2 - v^2}} = t, \quad x' = 0.$$

Thus

$$\eta = a\frac{c}{\sqrt{c^2-v^2}}y \text{ and } \zeta = a\frac{c}{\sqrt{c^2-v^2}}z.$$

Substituting for x' its value, we obtain

$$\tau = \phi(v)\beta(t - vx/c^2),$$
$$\xi = \phi(v)\beta(x - vt),$$
$$\eta = \phi(v)y,$$
$$\zeta = \phi(v)z,$$

where

$$\beta = \frac{1}{\sqrt{1-v^2/c^2}},$$

and ϕ is an as yet unknown function of v. If no assumption whatever be made as to the initial position of the moving system and as to the zero point of τ, an additive constant is to be placed on the right side of each of these equations.

We now have to prove that any ray of light, measured in the moving system, is propagated with the velocity c, if, as we have assumed, this is the case in the stationary system; for we have not as yet furnished the proof that the principle of the constancy of the velocity of light is compatible with the principle of relativity.

At the time $t = \tau = 0$, when the origin of the co-ordinates is common to the two systems, let a spherical wave be emitted therefrom, and be propagated with the velocity c in system K. If (x, y, z) be a point just attained by this wave, then

$x^2+y^2+z^2=c^2t^2$.

Transforming this equation with the aid of our equations of transformation we obtain after a simple calculation

$$\xi^2 + \eta^2 + \zeta^2 = c^2\tau^2.$$

The wave under consideration is therefore no less a spherical wave with velocity of propagation c when viewed in the moving system. This shows that our two fundamental principles are compatible.[5]

In the equations of transformation which have been developed there enters an unknown function ϕ of v, which we will now determine.

For this purpose we introduce a third system of co-ordinates K', which relatively to the system k is in a state of parallel translatory motion parallel to the axis of , [*1] such that the origin of co-ordinates of system K', moves with velocity $-v$ on the axis of . At the time $t=0$ let all three origins coincide, and when $t=x=y=z=0$ let the time t' of the system K' be zero. We call the co-ordinates, measured in the system K', x', y', z', and by a twofold application of our equations of transformation we obtain

$$t' = \phi(-v)\beta(-v)(\tau + v\xi/c^2) = \phi(v)\phi(-v)t,$$
$$x' = \phi(-v)\beta(-v)(\xi + v\tau) = \phi(v)\phi(-v)x,$$
$$y' = \phi(-v)\eta = \phi(v)\phi(-v)y,$$
$$z' = \phi(-v)\zeta = \phi(v)\phi(-v)z.$$

Since the relations between x', y', z' and x, y, z do not contain the time t, the systems K and K' are at rest with respect to one another, and it is clear that the transformation from K to K' must be the identical transformation. Thus

$$\phi(v)\phi(-v) = 1.$$

We now inquire into the signification of $\phi(v)$. We give our attention to that part of the axis of Y of system k which lies between $\xi=0, \eta=0, \zeta=0$ and $\xi=0, \eta=l, \zeta=0$. This part of the axis of Y is a rod moving perpendicularly to its axis with velocity v relatively to system K. Its ends possess in K the co-ordinates

$$x_1 = vt, \quad y_1 = \frac{l}{\phi(v)}, \quad z_1 = 0$$

and

$$x_2 = vt, \quad y_2 = 0, \quad z_2 = 0.$$

The length of the rod measured in K is therefore $l/\phi(v)$; and this gives us the meaning of the function $\phi(v)$. From reasons of symmetry it is now evident that the length of a given rod moving perpendicularly to its axis, measured in the stationary system, must depend only on the velocity and not on the direction and the sense of the motion. The length of the moving rod measured in the stationary system does not change, therefore, if v and $-v$ are interchanged. Hence follows that $l/\phi(v) = l/\phi(-v)$, or

$$\phi(v) = \phi(-v).$$

It follows from this relation and the one previously found that $\phi(v) = 1$, so that the transformation equations which have been found become

$$\tau = \beta(t - vx/c^2),$$
$$\xi = \beta(x - vt),$$
$$\eta = y,$$
$$\zeta = z,$$

where

$$\beta = 1/\sqrt{1 - v^2/c^2}.$$

......

SOME SIMPLE PRIOR KNOWLEDGE

Before the beginning of the next chapter, I would like to provide an introduction to physical laws related to the expression of natural laws. You can skip this part no, or return later when you want to understand clearer.

If your math is good, or you are not interested, then just give a glance to Figure 0.2. Later, we will have to frequently use these algebraic formulas (2) and related concepts. The rest can be skipped without reading.

You can also read only some of the text, ignoring those formulas. This will not affect your ability to read the whole book.

But, maybe, if you spend just a little time reading this section, you will appreciate the beauty of what is included inside the simple math.

Two Basic Formulas to Be Used Later

The first formula is:

Light Observed = Effect (Object Light Intensity, Distance)   (1)

We use (1) to represent the formula, and when (1) is used, it is meant to include the complete formula. It is a very simple description.

The formula provides a generic description of nature that is very abstract. In (1), "Effect" tells us that "the results obtained by the observer" are determined by "Object Light Intensity" and "Distance"; i.e., the results of the observer will change with variations in either the intensity of "object light" illumination or "distance." As for how the terms are specifically interrelated, we have to do the tests, theoretical summary, verification, etc. to get the law.

When using common mathematical symbols to express (1), Y can be used to represent "Light Observed," f can be used for "Effect," L for "Object Light Intensity," and R for "Distance," and the formula can be written as:

$$\underline{\text{Light Observed}} = \underline{\text{Effect}}\,(\underline{\text{Object Light Intensity}}, \underline{\text{Distance}})$$
$$Y = f\,(\quad L \quad,\quad R \quad)$$

That is

$$Y = f(L, R) \qquad (1.1)$$

Formula (1.1) is the same as (1). Expressed in language, it can be written thus: the light observed, Y, is the result of the function, f, worked on the object light intensity, L, and the distance, R, between the observer and the light source.

The next step is to find out what f is. This requires lots of research, experiments, calculations, etc. But in this example, scientists have already found the perfect law, which is:

$$Light.Observed = \frac{Celestial.Light.Intensity}{4\pi R^2} \qquad (2)$$

Replace (2) with common mathematical symbols, and we get:

$$Y = \frac{L}{4\pi R^2} \qquad (2.1)$$

We often write it in such way:

$$Y = f(L, R) = \frac{L}{4\pi R^2} \qquad (2.2)$$

Let's see the difference between (1) and (2).

Equation (1) qualitatively describes that observations will change with variations in light intensity and distance, but does not specify how the observations will change. Therefore, the two factors in parentheses are listed, but since the "Effect "of the law is not determined in the abstract representation, we only know that the results will vary in some way with changes in the factors in parentheses.

Equation (2) goes further. The "Effect" is defined by particular mathematical operators and the related parameters in the parentheses so that they are all connected together. The equation specifically tell us that, if you know the observational data of the light emitted by the target and know the distance between the target and the observer, then the light intensity can be determined.

The necessity for the second formula of limit theory can be understood from the interesting example of the tortoise and the rabbit.

The story is this: The tortoise is 16 meters in front of the hare and crawling at a speed of 1 meter per second. The rabbit runs at a rate of 2 meters per second to catch up. After 8 seconds, the rabbit reaches where the tortoise used to be, but the tortoise has moved another 8 meters, so he is still in front of the rabbit. The rabbit takes 4 seconds to cover the 8 meters, but the tortoise is now 4 meters in front of the rabbit again... No matter how many times the rabbit tries to catch up, the tortoise is always ahead, even though the rabbit is twice as fast as the tortoise.

In ancient China, similar problems have more concise descriptions.

In the Zhuangzi, in the Inner Chapters of the "world papers," it is written: "foot long stick, take one half per day, all ages inexhaustible."

Take a foot-long stick hammer divide it into two halves and take one half; the next day, divide that half into two and then take one half; the next day, do the same, and so on and so on... there is always half left. Over ten thousand years later, you can still take half; it is an endless task.

The tortoise and the rabbit story from ancient Greece and the foot-long stick hammer from ancient China, through similar reasoning, show that there are similarities in truth.

But in real life, the rabbit will need only 16 seconds to catch up with the tortoise. And in the foot-long stick hammer story, we won't be able to divide the stick in half after several days.

Is it meaningful to always strive for perfection given how absurd mathematical results can sometimes be?

Why do we get such absurd results when applying the perfection of mathematics to the real world?

This is a great challenge to human wisdom.

After a hundred years of arduous exploration, mathematicians established calculus based on the successes and failures of their theories. Limit theory is one of the core theoretical foundations of math and science.

Understanding Constantly Dividing the Foot-long Rod

Just as the foot-long rod can be divided constantly in half until it is, finally, not divisible, the light from a source will decline and finally disappear with increasing propagation distance from the transmission source to the receiver. This process can be represented by Figure 0.1.

We have to ask, if the distance R increases indefinitely, what will the final result be?

The answer is that, when R tends to infinity, the observable light L per unit area of a sphere of distance (radius) R from source C tends to zero.

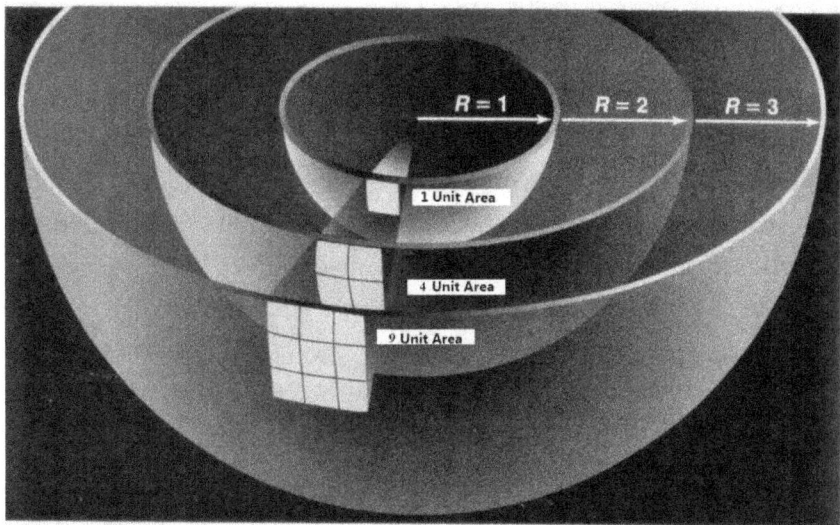

Figure 0.1 Observable light L sent from light source C with increasing distance R. As the area of the sphere surrounding C at distance R increases, the irradiation L per unit area of the sphere is reduced in proportion to the square of distance R from C.

The mathematical formula is written:

$$Light.Observed = \frac{Light.Emitted.from.Celestial.C}{4\pi R^2} \quad (2)$$

We would like to ask, if the distance R increases indefinitely, what will be the final result?

The answer is: when R tends to infinity, the limit of the light intensity per unit area on the sphere where the light source C is illuminated at a distance R is zero. Written as a mathematical formula is:

Light intensity per unit area from sphere of distance R of celestial C

$$= \lim_{R \to \infty} \frac{light.L.from.Celestial.C}{4\pi R^2} \quad (3)$$

$$= 0$$

This is a universal theorem that can be applied to a variety of light sources.

Let me explain to readers who have had no contact with limit theory. It is easy to understand.

| Debate of Light and Dark | Some Simple Prior Knowledge |

My eldest son is 8 years old. One day, on the way to school, I asked him a question: "How can you represent infinity?" He asked me what infinity means, and I said, "It is a number so large that there cannot be one larger. So how can you mathematically represent it?" I told him that if you add 1 to a number and it is still that number, then it is infinitely big. Of course, you want to use mathematical notation, which is ∞. The formula above uses the term "lim," which means "limit." The information written below "lim" indicates that the "lim" of distance R is infinite, or ∞. You can see from above proof that infinity is not a definite figure, but one that you can easily imagine "to what extent," so in rigorous mathematical language, we use the term "R tends to infinity" to describe it. Therefore, as the denominator R can increase to infinity, and the numerator is a definite figure, the final result of the whole formula is that it "tends to zero." Here, the most important thing is the dynamic thinking that the figure will change in a certain direction, which is the distinction between higher mathematics and elementary mathematics.

If the light source is a star, then the above limit tells us: no matter how bright the light is when emitted from the star, after the light propagates a certain distance, the light will totally disappear.

The Sun cannot shine to infinity. After a certain distance, the Sun will disappear.

This book will repeatedly apply the limit theorem of the propagation of light.

In the book we proved: once human telescopes are manufactured to a certain level, making better or advanced telescopes is useless. It will not help people to see farther. How far humans can see is not subject to the constraints of the quality of the telescope!

People are always growing up and advancing forward.

Space science, bounded by natural evolution, will sublimate its thought

AUTHOR INTRODUCTION

.According to a 1990 article published by The Daily Californian, the average time required for a student attending U.C. Berkeley's College of Engineering to earn his or her PhD degree is 5.9 years. In September of 1987, I enrolled in U.C. Berkeley to earn my PhD, which I did three years later in October 1990.

To complete this book, however, I needed over a decade.

Immediately following graduation from junior high in China, I was sent to the fields as a peasant due to the Cultural Revolution, where I spent the next ten years of my life planting, growing, and harvesting crops. As I toiled in the countryside, I taught myself China's high school curriculum, as well as university-level mathematics and English, developing my self-studying ability. After the Cultural Revolution, and passing the National College Entrance Exam, I was enrolled in college. As I earned my bachelor's from Shandong Mining Institute, my master's from China Mining Institute and Technology, and my PhD from U.C. Berkeley, I mastered the ability to think logically and became proficient with scientific research methods. This gave me the foundation necessary to complete "Debate of Light and Dark."

Throughout the ten years I spent working on "Debate of Light and Dark," I neglected enjoying life and would endure many sleepless nights in solitude, brainstorming with myself. It felt as if I simply would not be able to enjoy myself until the book was completed, as it was always in the back of my mind regardless of where I was or what I was doing. Though I learned to embrace the loneliness that accompanied this project, in fact, I really enjoy laughing with friends. My hobbies include: reading, writing, swimming, playing ping pong, playing go, playing the bamboo flute and tree leaves.

Published Works

- In 2005, I published Who Should Talk about Cosmos (谁有权谈论宇宙), a book from which one section, titled "Distance Mystique," was incorporated into the Chinese university national textbook.
- The essays "Hold American Dream" (托起美国梦) and "Gordian Title" (难解之题) both won World Daily Journal (世界日报周刊) literary prizes and were included in the Chinese anthologies Finding Dreams in North America (寻梦北美) and Studying in the US – Our Stories (留学美国-我们的故事).
- In 1988, I published "Computer Simulation of Management Operations Research" (管理运筹学-计算机模拟) in Chinese.
- In 1989, I co-published with a teacher the textbook Simulation Technique (模拟技术) in Chinese.

ENDING WORDS

I'm exploring the universe in pursuit of dreams,
Some happy, some anxious.
I would like to say thanks to this space for having no gravity,
So I was only confused, rolling,
Rather than stumbling over bruises.
I wanted to clear the fuzzy fog covering our eyes and ears,
Seize the touch of the golden light penetrating the dark,
But was surprised to find myself challenging the whole world,
Offending the past and present scientific gods.

Though living in such a bustling and lively information age,
I found my voice couldn't arouse pieces of echo.

But what's the big deal?
I firmly believe that time's flow will wash away all hypocrisy.
Truth is still shining after a hundred years.

It's a special surprise that you were able stick with the reading up to here.
Thank you.

www.ingramcontent.com/pod-product-compliance
Lightning Source LLC
Chambersburg PA
CBHW081606200526
45169CB00021B/2050